Author's Note

This effort was made for the purpose of bringing to a larger audience the author's view of an important but little known industry and its triumphs and travails - the American special machine tool industry. Hopefully that audience will include some who have been a part of that industry and who deserve whatever small measure of recognition this may provide.

It involves successes and failings of some inside the industry and of some associated with it from the outside. The intention has been to avoid inflicting discomfort on any organization or individual, but at the same time not soften any impact that may have resulted from their involvement.

After all is said and done, these are the author's views and opinions and not known to the author to be those of other individuals or organizations.

Acknowledgements

This effort has been profoundly influenced by industry entrepreneurs too many to list. They range from Eli Whitney to Edson Gaylord and Ralph Cross and their numerous contemporaries.

The following are just some of those who have also had an influence although in some cases they may not be even aware of it. They are in no particular order.

Bob Farris
Bram Bone
Tom Neff
Leo Weiland
Dave Williams
Melani Egbert
Harry Leaver
John Borseth
Bill Egbert
Hans Bretzner
Tom Powell
Trish Muehlenbeck
Bud Aspatore
Ron Compton
Tom Shifo

Vince Langely - whose use of some "reasons we work" and "TGIM" inspired their use in this effort.

And of course my wife, Barbara Egbert, who provided generous encouragement, sanity checks, and patience.

About Aspatore Books

Aspatore Books (www.Aspatore.com) is the largest and most exclusive publisher of C-Level executives (CEO, CFO, CTO, CMO, Partner) from the world's most respected companies. Aspatore annually publishes C-Level executives from over half the Global 500, top 250 professional services firms, law firms (MPs/Chairs), and other leading companies of all sizes. By focusing on publishing only C-Level executives, Aspatore provides professionals of all levels with proven business intelligence from industry insiders, rather than relying on the knowledge of unknown authors and analysts. Aspatore Books is committed to publishing a highly innovative line of business books, redefining and expanding the meaning of such books as indispensable resources for professionals of all levels. In addition to individual best-selling business titles, Aspatore Books publishes the following unique lines of business books and journals: Inside the Minds, Business Bibles, Bigwig Briefs, C-Level Business Review (Quarterly), Book Binders, ExecRecs, and The C-Level Test. Aspatore is a privately held company headquartered in Boston, Massachusetts, with employees around the world.

SWEET AND

SOUR

GRAPES

BY JIM EGBERT

ASPATORE
C-Level Business Intelligence™

Published by Aspatore, Inc.
For corrections, company/title updates, comments or any other inquiries please email info@aspatore.com.

First Printing, 2003
10 9 8 7 6 5 4 3 2

ISBN 1-58762-031-6

Cover design by Traci Whitney

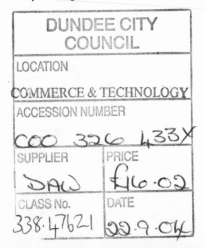

Table of Contents

Preface   9

Chapter 1: Imagination, Passion, and The Sixth Priority   13
            Imagination   13
            Passion   17
            The Sixth Priority   18

Chapter 2: Progress and Its Cause   21
            Progress   21
            Prehistory   24
            Natural Selection   24
            The Cause of Progress   27
            A Conundrum   27
            Value   30
            The Rate of Advance   32

Chapter 3: The Culture of Yankee Ingenuity   35
            The Genesis   35
            Maturity   40
            A Special Machine Tool   40
            The Payoff   46

Chapter 4: Why We Work (W3)   53
            Life 101   53
            Why We Work (W3)   57
            Recognition   58
            Classic Culture Clash   62
            Yankee Ingenuity at the Organization Level   67
            Seniority and Loyalty   68
            Job Security   70

Chapter 5: Historical Snapshot 73

    The Impact of WWII 73

    Evolution 77

    The Stress of Succession 81

    The Vitality Drain 84

Chapter 6: Leaders: The Facilitators of W3 87

    Business Management 87

    Succession Thoughts 92

    Recruiting 93

    Leadership 96

    Collar Color 100

    Courage 103

    Communicating 104

Chapter 7: The Business Enterprise 107

    The Special Machine Tool Enterprise 107

    Enterprise Personality 108

    Competing 113

    Ownership Examples 115

    Ownership Model 116

Chapter 8: Buyers and Users 119

    The Products Produced 119

    The Automotive Customer 121

    Procurement Practices 125

    The Auction 126

    Payment Terms 127

    Consolidation and Relationships 128

    Standardization 130

Personnel   132

Chapter 9: Competition   135

    Company Competition at the Level of the Individual   135

    Market Distortions   137

    Evaluating the Offering   139

    The Supplier's View   141

    The International Marketplace   143

Chapter 10: Government and Pig Farmer: 147

    Loan Guarantee Subsidies   148

    Tax Abatements   149

    The Special Machine Tool Industry and Government   151

    Monopoly   154

    The National Medal of Technology   157

Chapter 11: Lessons Learned   159

    Situation Summarized   159

    Lesson 1: The Global Free Market   165

    Lesson 2: The Human Element   166

    Lesson 3: The Business Environment   168

    Lesson 4: Managers and Owners   169

    The Buyers/Users Role   169

    Progressive Payments   170

    Standardization   171

    Technology on the Production Floor   172

    Lesson 5: Procurement and Use   174

    The Government's Role   176

    Lesson 6: Governmental Relations   177

One Final Lesson   177

End of Story   178

Addenda: 181

Addendum 1:

Lean Production and Agility   181

Addendum 2:

Machining Centers vs. Special Machine Tools   185

Addendum 3:

Managing Production Equipment   191

Addendum 4:

History   194

Addendum 5:

The Slide   196

Addendum 6:

Notable American Special Machine Tool Companies   201

Preface

This book traces Yankee ingenuity to its origins and its early American practitioners. Not realizing that they were engaging in a second revolution and establishing a unique new culture, they responded to their environment by confronting challenge and risk and satisfying their needs and wants. Examples involve prominent early Americans and resulted in the advent of special purpose machine tools.

There were characteristics common to these original practitioners that resulted in the rise of Yankee ingenuity. Imagination, passion, and courage would stand out in their resumes. Imagination, in fact, is a part of a definition of humankind – the ability to devise and use tools. We will see that all three characteristics are standard equipment for all of us, although they may be latent in many who require an awakening event or experience to enable them.

These characteristics are defined and chronicled and their impact on modern man and organizations explored. The freedoms fought and died for by the early Americans enabled their use and the development of Yankee ingenuity resulted. An acceleration of human progress followed, unlike that in any other society in human history.

Special machine tools are not broadly known or understood. Just enough knowledge is shared in this book to enable the reader to appreciate both the rewards experienced by the early Americans and the rewards we indirectly experience everyday. These rewards are the manifestation of continuously fresh intellectual products, the ideas, and engineering. They are cost and technologically effective machines that make production of manufactured products first possible, then practical and competitive. It's an industry that enables and fosters the Yankee ingenuity characteristics, advancing the quality of life for all.

The success or failure of every endeavor or enterprise results entirely from the effectiveness of the human elements involved. This book is intended to provide some added insight toward understanding that human element so that it may be more effectively utilized to the advantage of the reader and to stress the great value of one's life's work. It is particularly aimed at younger generations, who may not yet see their opportunities clearly, positively or at all.

Important byproducts of this industry are the promises of the reasons we work for all involved. They are powerful tools for individual and enterprise success. They are essential to the success of any organization in modern democratic society, though many do not recognize or understand that simple truth. The "Why We Work (W3)" logic is discussed in detail - the "sweet grapes."

Organized labor as we know it was an essential component of many industries into the mid-twentieth century. Its relevance in its current form in the global economy of the twenty-first century is questioned in this book. A modern competitive enterprise with an imaginative and passionate workforce expects and deserves W3 fulfillment, which inherently conflicts with certain organized labor tenets today. Management, in many cases, must likewise move to accept and facilitate that absolute of modernity.

Enterprise organization and leadership are discussed extensively. The premise that "the human element is everything" makes the challenge of an organization's environment an interesting one. Excessive rules, traditions, "that's the way we do things" paradigms, and mandated conformance to industry standards are repressive and counterproductive. Facilitation of W3 and the "find a way" culture of Yankee ingenuity are at the top of the "must have" list for successful companies.

The American special machine tool industry faces daunting economic and market challenges at the turn of the century. Many of these issues are typical of those confronting other industries as well. They are discussed and analyzed with prospective courses of action offered. Those actions will interest and benefit readers involved in many professions and vocations.

The principal marketplace for the special machine tool segment has been the global automobile industry and other vehicle and engine producers. The long-term growing strength of the American dollar and the loss of market share for American auto companies have seriously handicapped the American special machine tool industry.

The share loss translates to a comparable loss for that industry.

A problem in recent years has been procurement practices that sometimes put the initial low price paradigm ahead of real value considerations. In the case of special machine tools, it is sometimes done in an auction environment for the acquisition of their truly intellectual products. The discussion of these issues could be seen by some as "sour grapes."

The American special machine tool industry as we know it could be headed for extinction for reasons discussed. It would be a tragic loss for their customers and of course, for the industry's employees and shareholders. A major force in American history for advancing human progress and capitalizing on Yankee ingenuity like few others could disappear from the American landscape altogether.

Many of the thoughts expressed in this effort are simply reflections on a particular mix of lessons from the life experiences of a mature (older) person. The underlying belief is that the material, which may not seem as authoritative as academic text, is actually fairly logical stuff. Some of the thoughts, of course, are the opinions of the writer. It's reasonable to believe that lessons learned and opinions formed from a set of life's experiences could be beneficial to others.

Chapter 1:

Imagination, Passion, and the Sixth Priority

Imagination

Imagination: The great power of the human mind, which creates, associates, or alters mental images or ideas; the faculty of creativity.

Long ago, anthropologists concluded that the characteristic that distinguished human ancestors from all others, including similar species, was their ability to devise and use tools - the capacity of imagination.

"Rather than just using sticks and stones to leverage innate abilities - something done by plenty of animals from chimps to otters to finches - someone had deliberately selected raw materials in a sophisticated and consistent way with careful intent. This wasn't just tool use; it was technology."[1]

Does all this make you wonder if the oldest profession, in reality, wasn't that of a toolmaker?

Those tools were used for survival, for making difficult tasks easier, and to provide some level of comfort in the day-to-day lives of ancestral mankind. As time went on, their imaginations would conceive more and more things that would change their lives. It also gave our primitive

---

[1] Michael D. Lemonick and Andrea Dorfman, *Up From The Apes,* (Time Magazine, August 23, 1999), 57.

ancestors a purpose in life beside simply survival and procreation, as was the case in lower forms of life.

Their imaginations could make a better life for them and they could even begin to realize rewards beyond the obvious tangible ones. It's even conceivable that they could feel some level of elation at having solved a nagging problem with a new and unique idea, however primitive. They may even have received a primitive version of a high five or maybe just a knowing glance or smile from friends or relatives – recognition. It's possible that they also gained a measure of personal fulfillment in the process.

Can you imagine the feelings experienced when seeing the flint spark, a puff of smoke, and then a glow and finally, a flicker of flame for the very first time? Likewise, finding useful growth from the first seed purposely planted, mimicking Mother Nature's random wind blown method of sowing, surely elicited some special feelings as well. Human imagination had produced discoveries that would change lives forever. "The growth of the cognitive areas that distinguish ours from other large brains - could have come from our increasingly creative use of tools."[2]

Thousands of years later, that basic characteristic is stronger than ever in the makeup of modern man and it still serves the original purpose. The state of today's sophisticated and highly productive machine tools and the human progress that results from them is testimonial to this fact. Myriad other tools, services, and efforts such as computer hardware and software, medicine and agriculture, are also important.

In this book, an American machine tool industry segment whose genesis can be traced to a "Connecticut Yankee" is used as the basis for discussion of several subjects. It is the special purpose segment of the industry that is the focus of the discussions. It is an industry in which imagination manifests itself like few others, as its only products are a continuous stream of inventions. While not exclusive to the industry, the term Yankee ingenuity is used throughout the book and has been closely associated with it from their mutual beginnings.

---

[2] See note 1 above.

14

The industry that has been selected is unique in several ways. Why, then, is it used as an example for generalized discussion? Because all the challenges and stress normally associated with a new product cycle, from marketing and engineering to manufacturing and customer support, occur on nearly every order on a size and difficulty scale that could be fatal to the enterprise if proven unsuccessful. As a result, the various company functions are intensified in comparison to normal product companies, challenging all involved, and can be instructive.

This industry is just one example of those operating in an intensely competitive environment. The way that those pressures and challenges impact the organizations, the individuals involved, and the positives or negatives that result are all a part of these discussions. Many of these positives are not obviously tangible and require perspective to fully grasp. An objective of this book is to help provide that perspective. Additionally, numerous other industries and professions share many of this industry's characteristics and may benefit from the observations and examples that are discussed.

The subject industry, currently centered primarily in the Southeastern Michigan and Rockford, Illinois areas, is experiencing serious difficulty. Among the reasons are a market contraction, serious currency exchange rate disparities, questionable procurement tactics, and entrepreneurial, succession and estate planning shortsightedness.

With all due respect to New York City and its Yankees and the spirit that the team generates, this is not its story. However, in their vernacular, the global marketplace plays hardball and those expecting to succeed must have a competitive mix of imagination, passion, and courage to play.

Invent: To produce or contrive a new device, method or process by ingenuity and/or imagination.

The early Yankees, New England residents, sought independence from the Old World in all respects, including religion, liberty and self-determination. They especially wanted to be free from the environment of the Old World business monopolies and the culture of indenture. They wanted to be free to create the things that would make a difference in their time, an important plateau in the upward journey from primitive times to the present.

Indenture: A contract binding one person into the service of another for a specified term.

The new world environment was entirely different from that of the old. People were free, the needs of an exploring, pioneering, and settling population were altogether different, and an unprecedented abundance of natural resources was at their command. There were no precedents for the things that they could do. As with their primitive ancestors, they wanted more than just survival and procreation. They wanted a better life for themselves and their children and the other rewards that their liberty, passion, imagination, and courage would make possible through their own efforts - their "life's work."

Imagination is a characteristic common to all of us. Like others, it varies in magnitude and intensity. It is very likely that it is latent in many who have not yet recognized its potential or who have not yet discovered the passion and courage to experiment and develop that potential. The prospect of failure can be a daunting barrier to personal achievement, fulfillment, and progress.

Many of us do not think of ourselves in the context that in our time, we (all of us) are the source of the imagination and energy that create the things that make a difference. Those efforts can provide enormous personal fulfillment at the same time. It can involve material things, art, philosophical thought, music, medicine, agriculture, and countless other fields.

The human generations that live today are responsible for advances in such areas that would be incomprehensible to even recent previous generations. Advances continue to accelerate in following generations as the human element and its imagination continue to evolve and become even more effective. The individuals specifically responsible for those advances come from all walks of life, all levels of our democratic society. They have various skin and collar colors.

Passion

A layman's definition of normal human personality and behavior (character) would most likely involve three ingredients, whose proportions vary from individual to individual: family genetics, the chemical and electrical activities in the brain, and all of life's experiences. Those proportions then change with maturity.

Few of us can influence the first two. By definition, the third, life's experiences, is our opportunity to help shape people for success and happiness. It can be to mature in a way that utilizes the most beneficial influences, just as good grapes become a fine wine when their juice is aged in exactly the right cask material at the correct temperature, darkness, and time to an "excellent" maturity.

Early achievements, their rewards, and recognition for them excite the imagination to visualize even greater achievements, rewards, and recognition. This should not be seen as selfishness, since "rewards" go far beyond personal gain. In addition, passion may have been awakened in the process.

Passion (as in passion for one's life's work): Intense emotion compelling action; boundless enthusiasm.

Words alone are not nearly as effective as actual experience. Education, training, and especially responding to challenge and risk accelerate, broaden, and deepen the maturing process. Travel and exposure to diverse cultures can have a significant influence as well. A challenging "life's work" environment and related experiences enrich the lives of people and are an important positive influence on personality and behavior. A "life's work" environment that is empty and full of drudgery has the opposite effect.

Challenging: Demanding physical or psychological effort of a stimulating kind.

Some believe that at birth, life experiences while in the womb have already influenced behavior. Obviously, a one-year-old and a 70- year-old are on the opposite ends of the influences of life experience. With age, the other two ingredients are reduced in proportion and then personality and behavior are influenced largely by life's experiences. There is little being done at age 70 that doesn't have major influence from life's experiences.

Wisdom: Accumulated knowledge of life, which has been gained through experience, needed to make sensible, judicious decisions and judgments.

The special machine tool industry is uncommonly dependent on all of its people, their passion, and their imagination and courage to succeed.

Some important questions that arise in this discussion that may not be answered to the satisfaction of the reader: In which of the three ingredients of human character does passion, as in passion for life's work, originate? Can it be taught or influenced by life's experiences and can it be acquired through training or mentoring?

The Sixth Priority

It remains to be seen whether this effort can have any influence in helping to shape a reader or two. Interestingly, it has already affected the writer in the process, by lessons now recalled but not in sharp focus or not fully appreciated during the actual experience. It has sharpened the focus on:

Life's Priorities:

1. Our own life, its health, its longevity, and its quality. (Without the first priority the others would be greatly diminished).

2. The life of our mates, their health, longevity, and quality.

3. The creation of our offspring and their life's health, longevity, and quality.

4. Our other family members and their life's health, longevity, and quality.

5. The concern for and the treatment of others.

6. Our life's work.
   It has profound impact on the other priorities, on all those around us, and even on the future of our race. It is responsible for all human "Progress."

"Following the terrorist attacks on September 11, 2001, Maya Angelou said that hope has replaced the horror of that sight.

> 'I can see in the acorn the oak tree,' Angelou said. 'I see the growth, the rebuilding, the restoring. I see that is the American psyche. There is so much we can draw understanding from. One of the lessons is the development of courage. Because without courage, you can't practice any of the other virtues consistently'."[3]

In the "acorns" of passion, imagination and courage can be seen the magnificent, flourishing oak forest of modern democratic mankind, its freedoms, and its collective life's work: its progress

---

[3] John Smyntek, *Poet Angelou Speaks Out,* (The Detroit Free Press, October 9, 2001).

Chapter 2:

Progress and Its Cause

Progress

Progress: The sources utilized for various definitions in this effort have failed to provide one that seemed powerful enough to use for this topic. The question is, can we describe "progress" in a truly meaningful way?

If we compare the conditions involving human well-being in Biblical times, say those at the year 0001 A.D. to those at the year 2001 A.D., we could say that the human race has progressed. That would be true in just about all aspects of human existence. Life expectancy and health in general, for example, certainly would qualify. In this exercise, we are talking about just two thousand years, out of hundreds of thousands of years, of human history.

If we could show progress graphically for this particular period, it would be almost a straight horizontal line for the first 1700 years, or 85 percent, of the total. For those countries that we classify as industrialized, the Industrial Revolution, starting around 1700, began that curve on a more serious upward slope. Even considering that, through the late 1800s, or about 95 percent of the total, it would still be a very flat curve.

Can you imagine what that curve must look like after the late 1800s to get us where we are? It is five percent of the total time being discussed and involves only five of the 100 human generations of that period. The contributions of people like Pasteur, Edison, Ford, the Wrights, Einstein, Carver, Knudsen, VonBraun, Salk, Dell, Case, and Gates have given us progress in that short time that is hard to put into perspective. There are countless others like you and I whose life's work contributions may not show up on the same scale as those people, but nonetheless, the contributions that have been made toward human progress are priceless when added to theirs.

The impact of the American Revolution on all human progress cannot be overstated. It achieved liberty and self-determination as well as democracy for its people. That achievement, in turn, created an environment that inspired and energized the culture of Yankee ingenuity. The democratization of much of the world that has followed has compounded the advances made possible by that achievement. The global free market was subsequently born, however limited in its scope in the beginning, with its potential energy soon becoming apparent.

That imaginary progress curve for a number of countries, especially those that are sometimes referred to as Third World countries, has been well below that of the industrialized countries. However, those curves will come closer together sooner or later meaning an even steeper curve for them as they catch up.

If we visualize that curve, we will see that it is rising dramatically and we will see that it involves all of us. It is a dynamic curve as progress for the human well-being is accelerating. It is a composite of the human well-being and represents what is normal in relation to our endeavors, our enterprises, and ourselves. If, in the things that we are associated with, we are somewhere near that curve, our report card would be marked with a "C" even though we may seem to have made dramatic progress.

The message, of course, is that we must progress in what we do at the rate of the curve or better or we will fall behind since the curve represents what is normal. By the same token, if we progress faster than the curve we have an advantage over those with whom we compete.

The progress curves for individual industries vary from the composite human progress curve based on their maturity in terms of their industry. Agriculture, which was just as important in Biblical times as now, must be very close to the composite human progress curve. In the early days, if a farmer fell behind in the way he did things it was not a big deal. He could easily catch up, if he wanted, as the curve over a few years did not move vertically very much. If he didn't make the effort he would simply remain at the lower level of existence.

Today that same delay in reacting to market changes or technology advances could be deadly for an enterprise or even an industry. The distance to the curve in just a short time could be nearly an impossible objective to meet. It is like falling off a train at 5 mph, versus falling off at 100 mph. Progress waits for no one.

The curves for industries that came into being along the way, like the auto industry in the late 1800s, have progress curves which differ from the base curve as initial progress is typically fairly dramatic. By 1914, there were 300 American car companies, few of which could manage the pace and stay close to the curve being established by aggressive competitor companies. Today there are two such companies, Ford Motor Company and General Motors Corporation. As evolving industries mature, their progress curves will basically coincide with the composite human progress curve.

The American auto industry, relatively mature by the 1960s, had become complacent and drifted below the progress curve for the global industry. As an entire industry, it was overtaken in product quality, manufacturing effectiveness, application of technology, and in its speed of new product to market. Transplanted Japanese automakers Toyota, Nissan, and Honda led the way and others followed. Almost overnight the domestic industry was at a competitive disadvantage even in its own home market. Those competitors exceeded the progress curve and gained a serious competitive advantage and a substantial and still growing advantageous market share position.

The American industry has struggled greatly to recover its position on that curve. It can never fully recover from that lapse in its vigilance without beating its competition to the other side of that curve. At present, the competition defines the curve.

Prehistory

"The development of symbolic thought and complex communication did nothing less than alter human evolution. For one thing, high tech transportation means that the world, though ethnically diverse, now really consists of a single, huge population. 'Everything we know about evolution suggests that to get true innovation, you need small isolated populations,' says Tattersall, 'which is now unthinkable.'"

"Not only is a new human species next to impossible, but technology has essentially eliminated natural selection as well. During prehistory, only the fittest individuals and species survived to reproduce. Now strong and weak alike have access to medicine, food and shelter of unprecedented quality and abundance.' "Poor peasants in the Third World" says University of Michigan anthropologist Milford Wolpoff, "are better off than the Emperor of China was 1,000 years ago."[4]

This quote is both profound and thought-provoking and has implications ranging from religious and anthropological ones to global business issues as well. It is not possible to look at the business aspects in total isolation from the influence of the human side.

The true innovation referred to in the quote is the evolution of the small, isolated populations themselves that gave them the best chance to survive, reproduce, and prosper in their particular environment. Different environments produced different variations or peculiarities; in other words, different unique innovations. Meaningful human evolution producing those unique characteristics took many generations. The development of their local technology (tools and their use) was integral to that process and to natural selection.

Natural Selection

One of the implications of the quote is that human physical evolution, both now and in the future, is no longer directed by local environment.

---

[4]   Michael D. Remonick and Andrea Dorfman, *Up From The Apes,* (Time Magazine, August 23, 1999), 58.

Collective advances in medicine and in technology, facilitated by modern communications and transportation, now cause uniform evolution across the world populace. Even evolution of the human brain will become consistent across the entire world population with the interaction resulting from modern communication and travel.

Natural selection: The process in nature by which the plants and animals best adapted to their environment tend to survive and perpetuate the variations or peculiarities that enabled them to survive.

It is meaningful to examine direct consequences of those implications and to draw business parallels from them. The accelerating collective advances in medicine and technology have in effect replaced Mother Nature's natural selection process for the modern human. Those advances are the direct result of human imagination and energy - energized by free market competition - the modern counterpart of nature's natural selection process.

Today, natural selection is not competition for food or sustenance. It is the modern, global free market system that continually searches for better ways to provide greater value. The objective is to be rewarded with consumer business time after time by providing that greater value, beating the competition in the process and prospering as a result.

While the words traditionally associated with natural selection, "survival of the fittest," imply hurt for some, nothing could be further from the truth in modern natural selection. The successful competitor wins the battle at hand, but all humanity benefits from the largest to the smallest advances that result. All can enjoy the accelerating advances in modern medicine and technology. Life quality and life prolonging advances and abundance for all are the natural products of this modern natural selection process.

However, those enterprises that are not continually refreshing themselves, progressing at the rate of the curve and providing competitive value will not survive for long. The "survival of the fittest" phrase does apply to business enterprise in the modern natural selection process. The humans that are involved in that kind of situation, though put at a disadvantage temporarily, do have alternatives that will enable them to recover and to prosper.

In the year 1900, an American male could expect to live to be 46 and had almost no hope of retiring as we now conceptualize it. A woman could expect to survive to age 48. In the year 2001, those life expectancies were 74 and 79 respectively.[5] This is a 60 percent increase in life expectancy. The difference in only 100 years, little more than one lifetime, is astounding! Abundant retirements that last 20 or 30 years and longer are common. It's obvious that the preceding 95 generations of that period could not achieve the kind of life quality and longevity gains that are now possible. However, we get to live 28 or 31 years longer than those in 1900 and can expect a very nice retirement.

What will that be like for our children, grandchildren?

We have found the essence of a definition of progress that is powerful enough to use in this effort. More follows.

"As late as 1930, most American homes did not have a refrigerator, but, by the end of the decade, most did. By 1970 virtually all families living in poverty had refrigerators. By 1994, most American households below the poverty line had a microwave oven and a videocassette recorder - things that less than 1 percent of all American households had in 1971."

"Most American millionaires did not inherit their wealth, but created it themselves".

"Most people who were in the bottom 20 percent in 1975 were in the top 20 percent at some point before 1992. The poor will always be with us, so long as they are defined as the bottom 20 percent, even if yesterday's bottom 20 percent are now among "the rich" as such terms are defined by those with a stereotyped vision of a static world."[6]
Thomas Sowell describes this progress as "benefits of a free market."

---

[5] (National Vital Statistics Report, March 21,2002), Vol. 50, No.6.
[6] Thomas Sowell, *Free market creates social revolutions of past century,* (The Detroit News, January 2, 2000).

The Cause of Progress

"The growth of the cognitive areas that distinguish ours from other large brains - could have come from our increasingly creative use of tools."[7]

The implication from this quote is that the cognitive areas of the human brain and its general mental capability will continue to evolve on the basis of its creative use and exercise. It stands to reason, then, that competitors ranging freely, exercising their imagination in competitive combat stimulate that cognitive area's development and growth.

Nature's natural selection process took many, many generations to cause even small changes, while today advances in medicine and technology are accelerating geometrically. Natural selection has evolved to become the force that drives the small isolated populations - the individual enterprises of today, to create innovative products, processes, and medicines at accelerating speed - the free market forces.

The truly effective organizations are independent of and competitive with each other. They are small enough, at least in spirit, to foster an internal environment that nurtures passion, imagination and the courage to face challenge and risk. They are unencumbered by oppressive traditions, standards, paradigms, paralyzing bureaucracy such as government or industry, as well as organized labor imposed- guidelines, rules, or regulations.

A Conundrum

While modern communication and travel have untold benefits, they bring the world closer together in a way that can discourage the small, isolated population effect and competition - the conundrum.

There can be great temptation to do as others who are succeeding do and to standardize methods and processes across related divisions, companies, and even industries and countries. Consolidation of

---

[7] Michael D. Remonick and Andrea Dorfman, *Up From The Apes,* (Time Magazine, August 23, 1999), 57.

companies and industries also contributes to the equivalent of the single huge population. The modern natural selection process, progress due to free enterprise competition, is threatened by those trends. Nature's natural selection process was overtaken by the "single huge population" and became ineffective in much the same way.

For example: A pattern has developed where some in the automobile industry are spinning off their captive component parts production operations. Some of those new enterprises, in turn, acquire or consolidate with others. The smaller parts companies are disappearing. The plan is to sell components to as many auto companies as possible on long term mutual commitments. That arrangement will drive their volumes up and spread costs, including product development costs, over those larger volumes, in turn, lowering individual part costs.

The problem is that the costs are being reduced based on volume and by productivity improvements, at least in theory. Those lower costs are not necessarily the result of innovative, competitively-inspired products, better processes, and better methods and may not be unique to each of their customer companies. Competition, and therefore innovation, is reduced if not eliminated at the component level based on the sheer size of the producers, common production processes for their competing customers, and their preferred status with those customers.

Innovation by the auto part suppliers, in this example, may no longer be as prolific, since it is not necessary to beat all the competition to imaginative new products and better ways to produce those products. The larger component suppliers are not inspired to be aggressive, fast on their feet competitors. Flexibility and quickness are not necessarily high on their agenda. Huge entry barriers are erected in front of prospective smaller entrepreneurial enterprises, the small, isolated populations, discouraging fresh competition.

The short-sighted objective is high volume and resulting lower part cost with higher or assured long term profit. The competitively-inspired, innovative processes and methods that also result in lower part costs, along with continuous fresh new products for each final product, are not in the equation.

The supplier, by supplying competing auto companies, also narrows product differentiation and standardizes costs for all at the component level, ultimately diluting the innovation and cost benefits of competition at the final product level.

Prior to the serious entry of foreign automobile companies into the North American market, it seemed as though the American "Big Three," Ford, G..M., and Chrysler, had each other pretty well figured out. They knew what to expect from each other and generally accepted their respective market share position. In a way, it was a kind of large fraternity with few significant surprises model year after model year. Even their labor agreements mirrored each other and precluded any prospective labor relationship, cost, or flexibility advantage. From the inside, it would be hard to tell one company from another. If one began to slip in its position, it was likely because it fell asleep at the wheel, rather than being the result of an aggressive American competitor's challenge.

The baggage of tradition and labor restrictions produced a kind of apathy. Truly spirited competition between those companies was suppressed, and the promises of modern natural selection, free enterprise competition, weren't going to happen. The industry had become that single, huge population in the excerpt whose traditions, standard ways, and paradigms encouraged complacency.

The foreign entries, Toyota, Honda, and Nissan, were the small, isolated populations. They were unencumbered by the American auto company's paradigms, traditions, and labor restrictions, and they came with a spirit to innovate, qualitatively, technically, and organizationally. They came to compete. Those characteristics and the initial products were developed in their isolated home environment. Lean production methods, customer conveniences, fuel economy, and quality concepts are examples. (A well-known irony is that the quality concepts utilized were devised by Americans but rejected earlier as overkill.)

They were spirited, flexible, and aggressive competitors who accepted the challenge of breaching the local entry barriers, surviving and prospering in a hostile environment. It worked! Is more of the same imminent? As difficult as it may seem to accept the lessons learned from these experiences, at least some of them were addressed before they became terminal and in the long term those lessons will benefit all concerned.

"Despite a flurry of interesting new products coming from Detroit-based auto makers, Asian and European brands are going to continue to steal market share from them for the next five years, a recent study finds. The reason is because the foreign vehicles are more exciting and appear to offer better value to consumers."[8]

Value

Value: The perception of worth in usefulness or importance to the possessor, in utility and quality.

This is a static definition. There is another component of value that relates to time and progress. You bought the best mousetrap that money could buy based on the traditional definition. You then found out that a new and superior concept had just become available from a competitor. You did not buy the best value. So while appearances say one thing, the real value may be something different based on the dynamic progress curve. Both the supplier and the buyer have a problem in that example for obvious reasons.

The three-year-old PC that is being used to record this effort cost double that of a device with the capability five times as great at the time this is being written. This may sound like a complaint, though actually it is a testimonial to progress. The buyer, the supplier, and the competition must try to anticipate that rate of advance to be able to decide on their most advantageous direction and what represents the best value, all things considered.

If traditional mousetraps or auto components are planned by a supplier because retooling their processes and equipment is cost and time prohibitive due to its high volumes, it is not a good plan. An intelligent

---

[8] Drew Winter, *Study: Detroit to lose share for five years,* (Wards Auto.com, January 10, 2002).

competitive posture regarding the industry's next generation product timing should forge that plan. A "fast on his feet" competitor may be ready to introduce innovative new component versions on lower production scales for his competition-minded customers.

The higher the volume, the more difficult and expensive it is to be fast on your feet for competition and market responsiveness and to take advantage of technological advances. However, there are important examples where a marriage of dedication and flexibility is the wise choice (see addenda.)

Value (author's "dynamic value" definition): The perception of worth in usefulness or importance to the possessor, in utility, quality, and innovation.

Innovation is a component of value in the same definition as quality? It's true, because passionate, imaginative, and courageous people manage the risks associated with applying emerging technology and gain an edge against aggressive competitors. He or she "found a way." This will enhance the utility component of value and the useful life of the subject over competing products. It can even make competing products or offerings obsolete.

This definition contains a reminder to some, whose paradigms may narrow their vision, that "quality" is only one of three components of value. Even the highest level of quality has little meaning in a product made obsolete by competitive innovation. Some say quality is the price of admission to a competitive exercise.

Is the direction of high-volume, large organizations supplying multiple auto manufacturers cheap and profitable parts the right one? It's probably the right one for the parts supplier, at least in the short- term, as he has a lock on nicely profitable business. But what about the auto company and the consumer, you and me and even the supplier, when value-conscious buyers become aware?

Smaller (maybe just in spirit) American suppliers or those competing in and from foreign markets could be faster on their feet and innovative. They could be better able to survive and provide each of their auto

31

company customers different, advanced, better, and cheaper products (dynamic value) - the small isolated populations. Remember that innovative methods and processes developed in a competitive environment can provide the product designer with versatility that he had lacked and that his competitors will still lack.

Enterprises and even entire industries can easily be blinded by their own paradigms. The perceived benefits of economies of large-scale parts production at the expense of real freedom to innovate and compete can be misleading. Restricting rather than providing the freedom to use their full resource of human imagination, passion and courage will retard their competitive posture against all comers especially the small isolated populations.

The Rate of Advance

The first steam engine, the Newcomen engine, was invented in 1702 and was first used to pump out mines. The first successful self-propelled road vehicle was a steam automobile invented in 1770 by Nicholas Joseph Cugnot, a French engineer. Karl Benz's first motorcar appeared in 1885. The practical use of potentially portable energy from its first exploitation for everyday transportation took nearly 200 years to develop, still a relatively short period of time in human history.

In 1903, Orville and Wilbur Wright were the first to achieve powered flight. In the early 1960s, the Blackbirds, the Air Force SR-71s, were flying at 80,000 feet in excess of 2,100 miles per hour, twice the speed of a bullet from a .357 Magnum, where their titanium skin reached 750 degrees Fahrenheit. The flight of the first Boeing 747 occurred in the late 1960s. Neil Armstrong walked the surface of the moon in 1969. It is difficult to put into perspective the difference in complexity of these transportation events and their development time. 66 years, less than one's life expectancy today, from Kitty Hawk to the Sea of Tranquility was "One giant leap for mankind," in Armstrong's famous words.

Prior to the early 1950s, the vacuum tube was the heart of all electronic devices. A late small version of the vacuum tube measured about ¾ of an inch in diameter and was about three inches long. The transistor, the successor of the vacuum tube, was invented in 1948. A single computer chip of today, the size of a postage stamp, can contain the functionality of hundreds of thousands of vacuum tubes or transistors - the integrated circuit. That chip can process computer code from every day language instructions at rates that could not be comprehended just a short time earlier. We all know of the gigantic advances facilitated by the microprocessor chips and computers.

The curve is getting steeper and steeper.

Even now, we are on other technological thresholds. Nanotechnology promises the miniaturization of miniaturization in electronics and much, much more in the relatively near future. The promises of biotechnology, the discoveries in the genetic code, the human genome, and the prospects of stem cell potential are limitless, albeit with challenges regarding religious and other implications.

These examples demonstrate the acceleration of advances in those areas that are most easily recognized by us today. Comparable advances are occurring in most fields of endeavor at the same kind of astonishing rates.

Global free enterprise competition, modern natural selection, is fundamental to continued and accelerating progress for human well-being.

Fair competition between even dissimilar enterprises can be very productive. Consider the competition between the sixteen-nation consortium team headed by a University Of Michigan scientist, Francis Collins, against an independent American company Celera Genomics Corp. The spirit of competition in both organizations accelerated discoveries about the human genome that have dramatically advanced the prospects for health and longevity gains for all and in the process, have validated each other's discoveries. John F. Kennedy rallied the American ingenuity energy when he challenged it to beat the Soviets to the moon.

We don't always think about the benefits of competition. For example, when passing through the camera aperture of a sophisticated modern scanning or imaging device to evaluate a life threatening body illness symptom, do we recognize the manufacturer's name on the device? Is it an American company or is it a German or a Japanese company whose imagination and passion produced the value/technological edge and was rewarded with the purchase order for the device? What must the losers do to be competitive the next time, as they must be, or face serious consequences?

In taking daily medication or to be immunized against possible crippling or deadly illnesses, do we think about the various companies who have won that business through their research and the imaginations and passion of the people in their organizations? What ever happened to diphtheria, polio, tuberculosis, small pox, and measles?

The definition of progress and, more importantly, the cause of progress has finally become apparent.

Progress (author's definition): Life quality and life prolonging advances (technology and medicine) for all.

The Cause: The natural result of human imagination, passion, and courage energized by competitive business enterprise. The fittest enterprises survive and prosper by providing imaginative, winning value in continuous competitive exercise. That process itself nourishes and develops the cognitive human brain thereby accelerating a magnificent benevolent cycle.

Chapter 3:

The Culture of Yankee Ingenuity

The Genesis

Yankee: In the beginning, a native or inhabitant of New England; later a native or inhabitant of a Northern U.S. state; and finally and currently, a native or an inhabitant of The United States- an American.

In 1776, in Rhode Island, Jeremiah Wilkinson devised a new way to produce nails utilizing jigs and fixtures, the forerunners of special machine tools. In doing so, he was among the first to break one of numerous British monopolies. That monopoly by itself utilized upwards of 50,000 young, indentured servants making nails "the way nails are supposed to be made."

Paul Revere of Boston, Massachusetts and Midnight Ride fame, was not only a well-known coppersmith and silversmith, but also a manufacturer of ship's gear. He was also among the first Americans to succeed against the monopolistic metal working industries of England. He is less well-known for the saving of the *USS Constitution*, "Old Ironsides," an unseaworthy vessel, by his ingenuity in rolling a copper sheathing for its bottom.

In 1793, Eli Whitney of New Haven, Connecticut, responded to a challenge from cotton growers in Savannah, Georgia. The task was to devise an effective way to clean Upland Cotton of its seed. The invention of the cotton gin followed. In Whitney's words, "It makes the labour fifty times less, without throwing any class of people out of business."[9] It revolutionized the textile industry and started the South and the United States on a path of great domestic and international commercial success.

Whitney's real contribution however, was to manufacturing and mass production. Today, his 1798 success in advancing from jigs and fixtures to a machine

This was the ancestor of the special machine tool. It was a machine designed for a specific manufacturing step on one component part only.

Previously, all mechanical devices were assembled by "fitters," craftsmen who custom-shaped each piece to fit into its position in the device. No two of the pieces or of the devices themselves would be exactly alike. The way had been found to manufacture precision interchangeable parts, a concept previously thought not to be practical. You can imagine the complications imposed by the lack of interchangeability when thinking about the frequent repairs required by devices like firearms. It took a scarce fitter to make a new part to replace the worn part. It was said that at any given time, there were as many muskets waiting for repair as there were in service.

Whitney designed and built a "manufactory" and all the special purpose machine tools himself to make the interchangeable musket components. His machines would first produce 10,000 army muskets on a U.S. government contract. Whitney's revolutionary concept was the key to and the foundation of mass production.

Author Christy Borth describes that "major achievement as one from which virtually every item in our 'more abundant life' directly descends."[10]

---

[9] Constance Mcl. Green, edited by Oscar Handlin, *Eli Whitney and the Birth of American Technology*, (Library of American Biography, 1956), 46.

that repeated the same motion over and over again is not common knowledge.

[10] Christy Borth, *Masters of Mass Production*, (The Bobbs – Merrill Company, 1945) 27.

In 1818, Whitney invented the milling machine. Some believe his work to be the basis for the term Yankee ingenuity.

It is difficult for us to project ourselves back in time and into Whitney's position, but it is apparent that the courage he displayed in this undertaking was monumental. He had negotiated the musket contract with the U.S. government strictly on good faith, with no tangible proof that his concepts would work and with no previous experience in arms manufacture. The concept of interchangeable parts to most at that time would have been inconceivable.

The risk of failure in such an undertaking, with all the obstacles and all that was at stake, had to be extraordinary.

The risk taken by the procuring officials was also very significant, but Whitney's vision of the benefits of interchangeability was compelling and made the risk of failure worthwhile. The vision (imagination) and courage of those officials was fundamental to the success of those early Americans and we still benefit from it today. Those same characteristics are fundamental to progress in the twenty-first century.

Whitney demonstrated his concept to President John Adams, Thomas Jefferson, and other officials by assembling ten muskets from parts randomly selected by the officials from piles of the components that made up the assembly. The demonstration was required by the contract to satisfy the officials that the contract would be fulfilled.

The water-powered factory and the special purpose machine tools among the first ever to be used, had to first be conceived, then built and put into production. Whitney's personal fortune, his personal reputation, his workers' well-being, and of course the customer, a U.S. government preparing for war, were all at serious risk. He had the courage, energy, and passion to support his extraordinary imagination.

Note: While some controversy exists between various sources regarding the very first attempts at interchangeability, there is no disagreement that Whitney was the noteworthy American pioneer.

Still today, the courage to accept challenge and to risk failure is an essential ingredient for progress in any field. The risks must be borne by all involved, just as in Whitney's era.

Ingenuity: Inventive skill, imaginative and clever design and construction, having a clever and cunning mind.

These early Americans, "Yankees," and many others like them, were faced with monumental challenges in the new world. They were also blessed by having no guidelines, rules, traditional ways, standard methods, or "that's the way we do things" paradigms to suppress free thought.

There were no precedents for the things that needed to be done and that they would do. They were free to exercise their imagination and ingenuity. They would find imaginative ways to provide Americans and others the products that they needed and that would make their lives more comfortable. They would also give serious competition to the old world monopolies. They were passionate about their independence from the dictates of the old world and its monopolies. They were serious competitors.

A unique culture was born by virtue of the vast needs of a pioneering population, the competitive spirit, passion, and courage of those Yankees and "the no rules environment." This culture was no doubt grounded in the same characteristics that caused the early Americans to face the risks and challenges and to endure the hardships to make the hazardous voyage to America. The exploration and pioneering of the great American West and the journey to the moon were other examples of this same trait.

There will be more discussion on the "no rules environment" expression which purposely overstates the need for an atmosphere of freedom of thought and for acceptance. Any enterprise, of course, does require certain disciplines, common sense, and principled, ethical behavior.

Today, Yankee ingenuity applies to all fields of endeavor: manufacturing, medicine, agriculture, chemistry, the modern high tech fields of computers, information, and telecommunication and any others that can be imagined.

Yankee ingenuity (author's definition): The passionate, imaginative, and courageous search to find a way, "the solution," and then a better way and to beat the competition, whomever and where ever they may be, in the process. ("The best way" is never achievable as there will always be a better way.)

Formula: Satisfy apparent needs and wants, evolutionary or revolutionary, for apparent rewards - defined as the yield from the solutions themselves and "W3" – **Re. chapter IV page 41.**

While the term Yankee ingenuity signifies an American trait, it and the competing corresponding traits of other nationalities are "the cause of progress."

If necessity is the mother of invention, competition is its father.

The Yankee ingenuity culture grew along with the nation, expanding with each generation, and became a national characteristic. It caused the emerging United States' economy to accelerate and ultimately surpass the great powers of the old world.

American companies hold more than 90 percent of all global software patents, 90 percent of all global medical patents, and more than 80 percent of all global automobile patents. Many give the "find a way" Yankee ingenuity culture credit for winning the Second World War. It is largely responsible for creating the greatest economy that the world has ever known. The prosperity and comforts that Americans and many other nations enjoy entering the twenty-first century are the result.

Culture: Behavior patterns, arts, beliefs, institutions, and all other products of human thought and work; Intellectual and artistic activity and the works produced by it.

Maturity

Fast-forward two hundred years to the late 1970s and beyond. The American special machine tool industry is flourishing and the computer age is emerging at an astonishing pace

The fledgling industries commonly referred to as high tech, that is, computer and peripheral hardware products, software, and telecommunications, found themselves in the same blessed environment as the early Yankees were in. It was the "no rules environment" in a technological revolution that encouraged free thought and innovation. There were no precedents for things that needed to be done. Those industries would mushroom from nothing and prosper and the technology would advance at blinding speed in a culture of Yankee ingenuity.

The American special machine tool industry, the industry in which Yankee ingenuity manifested itself probably more than in most others, is at its peak. It is providing worldwide manufacturers a continuous stream of fresh ideas and the hardware and software to implement them. It was growing and prospering. .

In the mid 1970s, a manufacturer of small gasoline engines was preparing for a forecasted market expansion by evaluating costs and manufacturing effectiveness. In effect, he was preparing expanded production capability to be as competitive as possible in exploiting that potential. At the time, certain components, pistons for example, were produced in Asia for cost-efficiency. One of their new efforts included exploring the prospects of new and innovative manufacturing techniques wherever they might be found.

A Special Machine Tool

"Please provide your best proposal for equipment that will produce completely machined aluminum pistons, from permanent mold castings, ready for assembly into small gasoline engines at the rate of 360 per hour. The pistons are about the size of a large chickens egg and very

delicate as well. The production tolerances on several features are in the .0002" area, some features are irregular in shape and some require super finishing. The device is to include automatic loading and unloading (i.e. no operator). This device is being competitively bid and must be in production in a western U.S. state in 13 months from the date of a purchase order."

The above is greatly simplified, but is typical of a request for prospective special machine tool manufacturers to provide quotations to supply a unique process and its hardware for production of a client's product. This example is expressed in lay terms.

Special Machine Tool: (author's definition): Ideas and engineering manifested in cost and technologically effective processes and machines that make production of manufactured products first possible, then practical and then competitive.

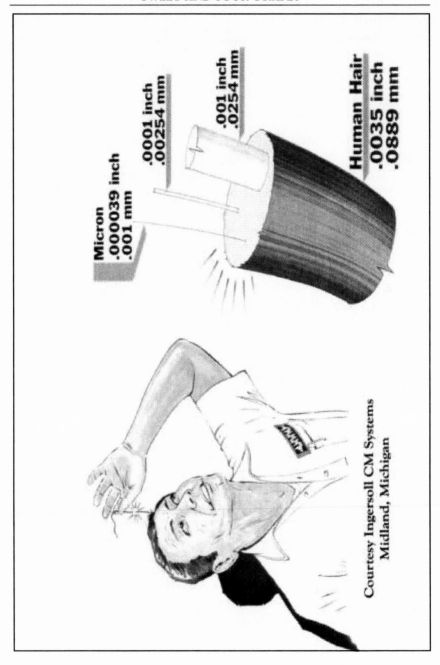

Micron
.000039 inch
.001 mm

.0001 inch
.00254 mm

.001 inch
.0254 mm

Human Hair
.0035 inch
.0889 mm

Courtesy Ingersoll CM Systems
Midland, Michigan

This is a request of a company, eventually assigned to an individual or individuals in that company, to invent something. Most of the time, it will be major machine tools and even entire manufacturing systems. It requires those people to compete, virtually head-to-head, with their counterparts among aggressive and clever competitors from around the world. It will be worth millions of dollars, apply the latest computer concepts with custom developed software, integrated with current and emerging mechanical technology. It will then be put into service on a very short schedule and must be supported in a location anywhere around the world.

A similar request is involved for any component to be produced in volume or those that would fall outside the capability of available standard devices. Equipment for machining of components for gasoline and diesel engines, transmissions, differentials, appliances, compressors, aircraft, and power generation equipment are very typical examples.

Consider:

- Conceptualizing an approach that wins a purchase order against clever worldwide competition to engineer and manufacture a $1 million to $60 million worth of unique, high technology, one-of-a-kind products.
- Refining the concept and engineering the equipment without time or budget for significant preliminary testing.
- Acquiring materials, manufacturing components, assembling, and demonstrating capability to the buyer. There is no tolerance for any equipment performance shortfall.
- Disassembling and shipping to facilities anywhere in the world.

Total elapsed time from receipt of purchase order = 10 to 18 months

- Re-assembling on production site and demonstrating capability.
- Training client operators and maintenance personnel and support early production.

To clarify, special purpose machine tools are basically metal cutting machine tools that can incorporate other kinds of operations providing additional effectiveness in a manufacturing system.

43

The point of this exercise is to include a discussion about the American special machine tool industry. It's an industry in which nearly every order is for a one of a kind product in a "take no prisoners" competitive environment. This mentality pushes the technology limits to offer the best value and win an order. Failure on a single order could be fatal for the enterprise and cause serious injury to the buyer. Success of that order, on the other hand, rewards both the buyer and seller. Eli Whitney would be like the kid in the candy store in this environment. The American special machine tool industry is little known to most Americans, but touches all of us in indirect ways.

In Europe and Japan, manufacturing and engineering are held in very high esteem. They are well-known and supported by industry, academia, banking, the government, their customers, and by the general public. In Germany, for example, it appears that the national psyche values industrial might or prowess on a par with financial might or other priorities, with both standard and special purpose machine tools in prominent positions. In the U.S., manufacturing industries such as machine tools and even automotive are no longer held in such high esteem and are sometimes considered to be part of a "rust belt."

American auto companies have discussed having a manufacturing presence in Japan beyond just joint ventures and are expected to move in that direction. That move is thought to be important strategically because manufacturing is so highly regarded in the Japanese culture. Many special and standard machine tool companies, including the second largest in the world, Amada, are located in Japan. Until recently Amada was the largest, but has now been overtaken by Comau, which is headquartered in Italy.

In the 1940s and 50s, many people believed that the auto and aircraft industries were mature and that significant advances to be realized had already been achieved. People today believe that very significant advances remain in both; in fact, the technologies are even accelerating. Even the gasoline and diesel internal combustion engine technology is still advancing significantly.

Alternative power sources for automobiles are now beginning to appear. Global positioning systems are evolving rapidly and are already in popular use. Guidance and crash avoidance systems will follow. Automobiles are being fitted with wireless communications gear that can interface with the Internet and drive-by-wire technology is being introduced.

Many types of technology must come together in an automobile and in an airplane in just the right way for those products to succeed and advance as they do. That will continue to happen, even accelerate. Many of the technologies involved are unique to each of the two industries, while others are more universal technologies such as computer hardware and software, but are uniquely applied. Even in the universal technology area, developments have been driven by the auto and aircraft industries and then applied in others. These products are really marvels of technology and new, more advanced generations appear with amazing regularity year after year.

The special machine tool industry as we know it was in its infancy in the 1940s, but it parallels the auto and aircraft industries regarding the development and application of technology. Previously unheard of speed in the movement of massive machine elements through use of linear motors, new concepts in absolute structural rigidity, and specialized computer hardware and software are among its advances. It is also not unusual for the industry to force technology advance for its own use from supporting industries as the auto and aircraft industries do.

The industry is an enabling industry for other manufacturing industries such as autos, aircraft, appliances, and others. It fosters advances in those industries by the development of new processes, and imaginative application of new technology. Heretofore impossible new product capability becomes first possible and then practical. Examples include: the ultra high strength fiber lay-up process and equipment for aircraft wing and tail structures; linear motor use in high speed machining of irregularly shaped highly efficient gasoline and diesel engine pistons; and the high speed machining from solid aluminum of gigantic wing spars.

It is comparable to the development of computer hardware and software by one industry versus its use by others. The high tech and special machine tool industries are similar in that they are both among the highly competitive and innovative technology originating industries, which drive technology advance - enabling other industries.

The Payoff

In the mid 1990s, an American auto company requested quotations from international special machine tool companies for several operations in a non-traditional approach to automotive crankshaft production processes. One of those had to do with drilling long oil passage holes. The traditional way involved a very expensive, many-stationed machine and, in some cases, two different machines in sequence. This type of machine is very specific for a specific crankshaft and they cannot be converted readily.

The time required to remove them from service for the conversion would be prohibitive in terms of lost production. In the traditional method, the engine designer cannot change anything in that engine that is affected by the location of those oil holes. That includes the stroke of the engine, which is a major factor in engine performance without major cost and time penalties.

The request was to find a way to make that operation flexible, providing capability to change the locations of the holes readily. This would untie the hands of the engine designer and would extend the life of the machinery through normal model changes, allowing processing more than one kind of crankshaft with simple changeover.

25 machines, each better than the last, have now been built for eight engine builders in six countries that accomplish all of that. They can produce at much faster rates as well. The American company whose solution was chosen had little or no experience in deep hole drilling, and therefore had no "this is the way it's done" paradigms. That provided a virtual "no rules" environment and Yankee ingenuity prevailed.

Those machines in turn inspired competing machines with similar attributes from around the world to keep those competitors in the game benefiting the procuring companies.

The piston example brought the production of these pistons back to the U.S. from the previous supplier in Asia, based on the innovative process and hardware proposed. The result for the buyer was a lower cost, higher-quality piston at higher production rates. It did not take long to return the investment at that production rate. Several more of these machines were built for competing producers, as well as a second machine for the first buyer, each better than the last.

In the wing spar example, the completed machine tool will do at least five times the work of any previous machine tool designed for similar purposes. That is high enough to justify a change in or a practical alternative to aircraft wing construction techniques. This is an important example of the buyer benefiting from manufacturing's technological advances, precipitated by an American special purpose machine tool company's vision of the future.

The challenge involved in the conception of this machine and then the commitment to implement it for the first time for a real live production requirement rivaled that of Eli Whitney's musket challenges.

The imaginative solution, competitive price, and delivery promise determined supplier selection. The risk is that the cost of failure or delivery delay could easily exceed the net worth of the supplying company. For example, the financial consequences of not producing the gasoline engines because of piston problems for an existing chainsaw market at 360 per hour, three shifts, and seven days a week is mind-boggling.

The courage to accept the challenges and take the risks evidenced by the producers of these special machine tools, like those taken by Eli Whitney 200 years earlier, is fundamental to the realization of the fruits of human imagination - progress.

The special machine tool engineers are a unique blend of manufacturing engineers and machine tool design engineers. First they devise the winning competitive manufacturing process solution. They then engineer a unique machine tool, hardware and software, applying contemporary and emerging technology for the production of component parts for other industries. They grow and prosper by finding new and better ways. They out-think their worldwide competition for every order, providing value and manufacturing advances to their customers. Supplying a quality, reliable, first of a kind, complex product that satisfies a discriminating and demanding customer from anywhere in the world, under extreme production pressure, is truly an adventure. Profiting from the experience and furthering the organization's reputation returns a great sense of accomplishment.

The notable special machine tool companies have each had staffs of a 100 or more engineers. In normal busy times, each company managed a large number of contract engineers as well. Examples exist of those engineers and others of their peers advancing to the highest level of responsibility in those companies. There are also significant examples of those who left those companies to begin their own successful related businesses, in some cases selling engineering capacity to the very companies they left as well as to competing companies.

During the 1960s, 70s, and 80s, the largest of the then-American companies, F. Jos. Lamb, Ingersoll Milling Machine Company, The Cross Company, Bendix Machine Tool Company, EX-CELL-O Corporation, and LaSalle Machine Tool Company competed for business from the global auto companies. They served the heavy vehicle, aircraft builders and other industries as well. The total number of people staffing their engineering departments, including their contractors, would probably exceed 1,000. They were typically paid for an average of 55 hours per week. 55,000 hours of specialized engineering every week were sold to their customers as a part of special machine tool orders for the advancement of their production capability - the intellectual portion of their products.

It may seem to be a meaningless extrapolation to some, but in the 30 years suggested above, there were in the neighborhood of 80 million hours of engineering experience accrued. Those hours advanced the state of the art for the production of machined parts largely for the American auto companies. The free enterprise principles of the day and Yankee ingenuity produced dynamic value by developing and exploiting emerging technology. The loss of that accrued experience, irreplaceable by itself, is not nearly as serious as the loss of the capability to continue and to exploit the now exploding technological goldmine.

That engineering capability was focused on the development of the new and unique concepts that would satisfy orders won against other clever manufacturing engineers from around the world. Approximately 20 percent of an order value was engineering content and consumed approximately 40 percent of its schedule.

That capability today is a very small fraction of what it was in those years and is still declining and could even disappear. One might get the idea that it must no longer be required. While there is no intent to directly connect the two, the American automotive market share in 2002 in the North American market is about half of what it was in the 1960s and is still declining. Logically, no potential advantage such as the use of that accrued experience and the Yankee ingenuity behind it should be overlooked.

There are times when we are so accustomed to our surroundings and our activities that some important things may not even register in our consciousness. As a society today, we are so preoccupied with an astounding array of high tech products and what is being done with them and what is expected from them in the future that we take for granted other less glamorous products.

The automobile is an example of this. It still has four wheels and an engine, but its present form would be totally foreign to Henry Ford or Karl Benz. It has become indispensable to the American way of life. Yet, its engineering and manufacture is thought to be part of the American rust belt even with the very high level of high tech content. Special machine tools are a close parallel.

49

If we lift our heads up and look around, it's obvious that most of the items referred to in Christy Borth's "more abundant life" are manufactured items. When we add consideration for the impact of newly available technologies, including vast computer application possibilities, Yankee ingenuity is more relevant in manufacturing today than ever before.

The special machine tool industry has been an important part of American industry and the advances fostered by the benevolent cycle (progress). Today the industry is seriously failing and its very survival is in question. There are several reasons some of which are discussed in the following chapters. Two prominent reasons follow:

1. The American dollar has steadily gained strength against the German and Japanese currencies since the late 1960s. For example, the deutsche mark was less than half its value against the dollar in 1998 from what it was in 1971. Today, the Euro has regained some strength but is still about 55 percent of what the deutsche mark was in the late 60s. The Japanese yen in 2001 is about a third of what it was against the dollar in 1971. Those relationships have a direct effect on the competitive position of American companies against the German companies, their main competition, and against the Japanese companies. The American dollar will buy about twice as many German machine tools today as it would in 1971.

    Some believe that Yankee ingenuity will offset part of the currency handicap but it is still a very serious problem.

2. The American auto companies (the principal market for special machine tools) had about 65 percent of the North American market in 2001, down from 85 percent in 1975. Both the auto imports and the transplants are very nationalistic when selecting capital equipment for their production needs, which makes auto market share a direct translation to the size of the market for machine tools.

Today, the real special machine tool market is 76 percent of its size, and foreign competition has a 2-1 price advantage (currency exchange rate) compared to that of 1970. Chrysler, now owned by Daimler Benz, has about 13 percent of the market. It's entirely conceivable that the corresponding special machine tool market share could follow.

These issues also affect American companies operating in other countries the same way. The local, transplanted, and imported auto companies in the European market for example utilize their own national suppliers and the currency exchange problem is duplicated as well.

Chapter 4:
Why We Work (W3)

Life 101

A recently-aired National Public Radio interview of a well-known and accomplished jazz musician covered his entire life. He talked about his passionate feelings for music, even when he was a preteen. His hunger for knowledge and for the understanding of everything that went into all aspects of music were like a fire in him. It continues today, and he is in his 60s. He became accomplished in the many different kinds of music, from the classics to jazz, blues, be bop, etc. He was passionate about his "life's work" from the beginning. Throughout his life, he used his love of music to make a living, a good living, although it was a struggle in the beginning.

How many of us earn our livelihood, "our life's work," doing that which we love most to do?

We spend about 75 percent of our adult years and about 50 percent of our awake hours during those years working for a livelihood. We will use the majority of the energy that we will expend in our lifetime. The point is, of course, that we had better like, love, have spirit about or even be passionate about what we choose to do for that livelihood, our life's work, as the jazz musician obviously did.

Little is more important in our lives. It is the sixth priority of our lives, only behind the sanctity of the lives of our loved ones and our other human relationship values. It is not altogether clear to many that the choice is really ours as individuals. It is our choice as members of a free and democratic society. Preparation for that choice should be more seriously taught, beginning in our homes and very early in grade school, and again and again, imbedded in school curriculums. The knowledge and understanding of the great value of our life's work, for the individual and for its part in human progress, is as important as any subject taught today. It is truly "Life 101" and a subject taught throughout life, if we are paying attention.

"Work does more than get us our living. It gets us our life."
--Henry Ford

A Common Riddle:

Count the F's in the following sentences:

FINISHED FILES ARE THE RESULT OF YEARS OF SCIENTIFIC STUDY COMBINED WITH THE EXPERIENCE OF YEARS

Done? Don't go on reading until you got them all!

How many?

3, 4, or 6? Try again.

A count of three is normal, four and five are not very common, and six is rare but is the answer. Our paradigms virtually blind us to the word "of" as it is not seen as a word but just a connecting device.

Paradigm: An example that serves as a model or pattern for following efforts.

This classic definition of paradigm does not convey the deeper meaning that is implied in its common usage today. The pattern or example that may be referred to can be so compelling that we are virtually blinded to real and altogether practical alternatives. In some instances, we actually cannot see very obvious competing and practical alternatives to long held beliefs and "that's the way things are done" rules. That blindness can and does seriously limit our horizons. It can be true for an individual, an enterprise, an entire industry, or even a country. The retrospective "what might have been," improves our vision, but usually when it is too late.

The 50,000 laborers making nails in England were blinded by their own paradigms, while Jeremiah Wilkinson had 20/20 vision since he likely had no idea how nails were "supposed to be made." In Eli Whitney's day, the idea of interchangeable parts just didn't register with most, as it was counterintuitive to their paradigms of the fitter or craftsman having to custom make such devices as firearms one at a time.

Our paradigms lead the career naval officer's son to a naval career, the coal miner's son to be a coal miner, and the farmer's son will likely become a farmer. These examples are not problems and may be less likely today than in earlier times. A paradigm problem can exist for those who are or perceive themselves to be disadvantaged. Some in environments that are distressed virtually cannot see outside their paradigms. They may believe that their lot in life is cast to follow their parents, no matter how loving or how disjointed they might be, or the pattern or model of their present environment no matter how harsh. Many seem to accept their present circumstance as their fate. A perverse comfort in knowing what to expect regardless of the severity, as opposed to the unknowns of change, no matter how optimistic, is part of the paradigm effect. They may be blinded to their real opportunities.

On a larger scale, some non-democratic societies have paradigms, probably defended by their leaders, that limit their opportunities for better lives. They may not enjoy the personal liberty that Americans do or the freedom to satisfy "the reasons we work" limiting their vision of a better future for themselves and their families.

Our default internal software (paradigms) has been programmed to resist change. Avoiding the risk of any loss of present perceived comfort, thus blinding us to alternatives, is part of that resistance. An important characteristic of those who have exploited Yankee ingenuity and of successful individuals in general, is their ability to reprogram themselves. They enable their vision. They can see clearly the existence of alternatives and their possibilities in all that they do. They have awakened and exercised their imaginations.

"Imagination is more important than knowledge." Albert Einstein

Maybe we should find a different term for our "life's work," since work can imply unpleasantness, drudgery, or flat-out labor. During our childhood, some of us who observed our parents and relatives struggling to realize some measure of prosperity may have come to regard our life's work only as labor. That expected unpleasantness was to be avoided whenever possible in favor of something that was fun to do. If life is to be truly meaningful, they must be one and the same.

Labor: Exertion, toil, drudgery - little or no connection in spirit to the outcome.

Work: To put forth physical and/or mental effort, to act effectively, - to accomplish the end desired, sometimes considered a work.

Note: The difference may only be in the attitude, spirit, passion or paradigms of the individuals involved.

A Work: Something accomplished through the effort. A creation, things that matter.

"Why don't TGIF (thank goodness it's Friday) and TGIM (thank goodness it's Monday) have equal importance?" The paradigms of many of us do not allow TGIM to even be comprehended, let alone be used in the same sentence as TGIF. It must be sacrilegious.

"According to Aristotle, 'Every human seeks happiness.' Morris said, "There are three fundamental views of happiness; Pleasure, personal peace, and participation in something that brings fulfillment."[11]

---

[11] Tom Morris, "*Know thyself,' philosopher tells Matrix audience"*,(Midland Daily News, May 22, 1999).

A significant amount of space could be devoted to discussion about the certainty that there must be balance with quality family time. The top six priorities of life, human relationships, and our life's work must complement each other and be woven together into the exquisite fabric that is our life.

Why We Work (W3)
(The Sweet Grapes)

Why do we work?
- For recognition
- To find solutions for apparent needs and wants
- For feelings of accomplishment (personal achievement)
- For fulfillment
- For security (more than financial)
- For team competition (as in athletics)
- For enjoyment
- For livelihood - family prosperity
- For companionship (camaraderie)
- To be a part of something
- To gain skill and wisdom as we mature
- To experience two way teaching and mentoring
- For self esteem (dignity)
- To be the best we can be
- To feel ownership and responsibility
- For the need to be vital
- For independence (liberty)
- For self determination
- To experience success and failure (to mature)
- To make a difference (make a mark)
- For pride
- For self expression (creativity)
- To be challenged
- For peace of mind
- To create the things that make a difference in our time
- To leave a legacy of things that matter

- To exercise all human functionality for it's growth, health and longevity
- For the fulfillment of our dreams

Are these also not the things that set us apart from lower forms of life, things which cannot be given, purchased, legislated, or bequeathed, regardless of the heights of our affluence or the depths of our poverty? While it may not be thought of as an individual reward, isn't progress for all humanity the ultimate reward for all of us? What would life today be like without our ancestors' W3 mentality?

W3 is the chemical formula for Yankee ingenuity and for progress. It's all healthy stuff and if we think about it, the best part is that achieving fulfillment enhances financial well-being as well. It appears that the musician experiences all of the above. Why shouldn't we? Nothing could be more natural or logical. Shouldn't our upbringing, particularly our formal schooling, focus the attention of youngsters on this, one of the most important lessons of life itself? It appears that many times it is approached very casually both in the home and in school environments.

Any company's primary resource is its people. Realization of the promises of the "reasons they work" benefits the company and their coworkers as much as themselves. They become the most powerful attribute a company can have to succeed and prosper. It's a "turned on" organization in contrast to some that are in fact "turned off" and experience the tragic waste of day-to-day empty drudgery. The "turned on" organization provides the best possible job security and facilitates Yankee ingenuity in the best possible way.

Recognition

The first item on the list of the reasons we work is there because it is the greatest natural motivator known, not just for management's ends, but for employee fulfillment as well. It is true for all of us and even for lower forms of life. It can take many forms, but recognition is an expression of appreciation and many times with sincere good wishes for the accomplishments of another.

"Remember that a person's name is to that person the sweetest and most important sound in any language."[12] This is a form of recognition. Just saying someone's name can pick up his or her spirits just because you have singled them out.

A child having realized a success in a play or in a sporting event will likely hear "Good job." Athletes will often get a pat on the back when achieving a success from the coach or from teammates. They may offer a "high five." Our pets will get a scratch, affection, and maybe a treat when doing a trick or obeying when asked. Even a killer whale will get treats and affection when doing its tricks.

In business, recognition is common and can include things like a prime parking place for employee of the month, or prizes like golf clubs or a trip to Hawaii.

There are bonuses or other forms of incentive or merit compensation, like time off with pay. There are companies where a significant amount of normal income is derived from incentive compensation, recognition for having achieved certain objectives.

Recognition is a feel-good thing for all of us, whether we are giving it or receiving it. It helps to cement team spirit, it can help to take a load off, and it will lift the spirits when other things cannot. Many times just words or a touch are enough to make a difference. Other times, more serious recognition makes sense, particularly when an effort and the following success go on for some period of time.

Organized labor will not negotiate recognition in the form of merit compensation for their members in labor agreements, since it violates their basic tenet of "equal pay for equal work." In addition, their philosophy implicitly discourages recognition in any form as it is perceived as "management's prejudices and preferences" and it is apparently seen as lowering the rest instead of raising the ones being recognized.

---

[12] Dale Carnegie, *How to Win Friends and Influence people* (Pocket Books, 1998), 79.

Many of "the reasons we work" listed have to do with personal achievement. When good things happen, people notice and natural recognition occurs. It is unnatural to suppress the inner or outer celebrations when positive things happen, and a significant, maybe unconscious, letdown occurs when nothing is said or done. This can be de-motivating.

In any ongoing effort to achieve goals, whether self-imposed or those assigned by others, frequent small achievements or victories manifested in recognition are an important feedback system. They provide vital encouragement to sustain the effort to the final success. Without the feedback, we are virtually adrift in unknown waters. Recognition is that feedback that provides encouragement and the incentive and direction for further action. This is particularly true for young people who inevitably face unfamiliar territory and are experiencing life's adventures for the first time. It is a way of subconsciously measuring one's own efforts and progress against the original objective and society's norms.

The following is an excerpt from the Internet home page of an international union as it entered contract negotiations with major automakers.

"To that end, we set forth the following program:

- Opposing new "two-tier" arrangements and seeking to eliminate those that currently exist. Two-tier wage systems violate the principle of equal pay for equal work, pit senior workers against new hires, have a corrosive effect on worker morale, damage productivity and create second class citizens in the workplace. For all of these reasons they must be strongly resisted.
- Eliminating wage progression schedules that require unreasonably long periods to reach the top rate for the job and thereby create, in effect a two-tier system.
- Replacing merit rating and similar wage progression systems with full automatic wage progression to the

top rate for the job. The pay of our members should not be subject to management's prejudices or personal preferences. Most of these "systems" exist in new bargaining units which maintained such wage payment plans before their workers joined our union."[13]

The principle of "equal pay for equal work" as described in this doctrine can easily be interpreted to mean that all must gravitate to the lowest common denominator so that no one can be seen as a second class citizen.

"Management's prejudices or personal preferences" is code for recognition of contribution beyond the minimums that have been specifically negotiated. It is not unusual for peer pressure in a closed organized labor environment to chastise peers as suck-ups for any contribution beyond that required. Of course, this further degrades spirit and effectiveness. The chastisers apparently become second class citizens when one of their peers makes such a contribution. It becomes a poisoned environment.

If we accept that Yankee ingenuity and the natural reward of the "reasons we work" are reasonable concepts, then consider how an environment containing the above philosophy contrasts with it. It runs exactly counter to the "no rules" environment, totally repressive to free thought and the energy for the creation and implementation of new ideas. It places huge obstacles in the way of the realization of the rewards of W3. The human element, its free thought (imagination) and energy (passion) and Yankee ingenuity are impotent in that setting. Big time TGIF.

In the event of another national emergency such as WWII, what will be the source of imagination and energy in manufacturing to perform the required miracles? What do you think about other countries and cultures, not only in terms of national security, but also as everyday competitors over the long haul? In today's global economy, very aggressive and ingenious competitors come from all points of the compass. They

---

[13] *UAW Bargaining '99 Economic issues* (UAW Online Publications)

seriously threaten job security for those not willing to take them on - on their terms - if necessary.

Michael Jordan is recognized around the world for his athletic ability, as he should be, as well as for his personal reputation. Do you suppose that the least well-known player on the same basketball team felt like a second class citizen because of it? What about the mission commander on a NASA space mission who gets all the recognition through his own organization and through the media? Do you think that all the others involved, including crew and support people, feel like second class citizens?

Classic Culture Clash

In the 1700s, machinery began to appear in mills and cloth factories. The objective was to reduce manpower (work content) as facilitated by technological advances, thereby reducing cost and increasing production and uniformity in the products. Those objectives were achieved. In fact, the human work content was reduced dramatically.

All competing mills had to respond in kind or face serious consequences. What a great accomplishment! The mills prospered, the buyers of the goods got lower prices and higher quality and an industry progressed to a new, higher level of efficiency. Everybody wins, right? Wrong! The mill owners did not feel as though they had any obligation to the workers beyond their pay, and they were fired with little hope to recover employment in their repressive society.

So what happened? The angry workers, called Luddites, responded by destroying machinery, burning factories, and probably worse. It was a primitive reaction, but it was probably their only way of being heard.

Luddite: One who opposes technological change. In early industry, Luddites rioted and destroyed labor saving machinery that they thought would diminish employment.

Were they able to stop progress or, in their eyes, the only thing that mattered, save their jobs? They did not and could not stop progress.

Secret societies were formed. They recruited all who might be affected, some by intimidation. Their purpose was to preserve their members' jobs by attempting to slow progress and technological advance and the resulting reduced work content. Governments eventually recognized those societies and labor unions were born.

If we could turn back the clock, do you think we could convince the mill owners that a way could be found to save those jobs? Maybe they could expand in other areas, or reduce the workforce through attrition over a period of time, accepting reduced profit in the meantime. Not a chance! Maybe we could find a way to stall that technological advance and the work content reduction indefinitely.

This is another conundrum: A paradox whereby progress was not possible without major disruption and hurt for the workers directly involved, even though there would be countless individuals and generations who would benefit greatly.

At the threshold of the twenty first century, nearly 300 years later, these same issues are important factors in the relationship between management and labor. While we cannot turn back the clock and fix a problem of the past and cannot stop progress in the future, we should be able to fix that problem for the future now.

To have a chance, we first must understand the problem and get it into a perspective that allows it to be viewed from all sides. Technological progress will always be a reality. It will even accelerate, and by definition, it reduces human work content. In viewing human progress over those 300 years, would anyone have it any other way? Consider future years for our children and grandchildren as well. Would we stop progress now if we could to avoid the problem for the future? Not a chance, and for the same reasons! Change is here to stay.

Since the days of the secret societies, slavery has been eliminated, world wars have been fought to protect and free entire races and populations, apartheid has been eliminated, and democracy prevails in much of the world. Liberty and self-determination are common. Humanity to people today, while a long way from perfect, is in a different arena than in those days.

But our conundrum is still there.

In a modern business environment, the workforce, those who actually perform the work content, likely has a good overall understanding of what is required to accomplish tasks. It will be from a different perspective from those directing it. Many times those same people will have thoughts on how to improve what they do and how they do it. In many cases they have not really been encouraged to make them known. In fact, the "suck up" mentality and its environment discourage it.

The $64,000 question is: Will they enthusiastically volunteer those thoughts and ideas with the notion that their work becomes more effective and helps all involved, including their company's competitive position, and thereby their own job security? Will they be able to realize the W3 rewards? Is it a "turned-off" organization, or is it "turned-on?"

Diagnosis: The harder we work to find better ways of doing our job, embracing new and more effective technology, reducing work content, and increasing product dynamic value, making our company more competitive (contributing to progress), the higher the likelihood that we will no longer be needed. Company success is supposed to create job security!

Naturally, this is a complete and total turnoff. Someone has it backwards. It is a reverse reward or recognition system and that is the basis of the problem in a nutshell.

Prescriptions: Business management has the responsibility for its organization to be at least as effective as its competition. That cannot happen without the enthusiastic contributions of its people, the human element. If you don't believe it, you need to understand what is going on in other companies and other countries - the competitors.

- The organization must be assured that their own contributions will benefit themselves in the improved company effectiveness and that job security will result for all involved, not the opposite. An atmosphere of mutual trust and one where the promises of the reasons we work can be fulfilled must prevail.

- Organized labor has a responsibility to protect their member's jobs. Without the threat of job loss due to reduced work content, job security can best be defined as success of the company and effectiveness at the employee level in the global marketplace.

- Company success (job security) is achieved by the enthusiastic contribution of all the people involved, the human element. That is best achieved in an atmosphere of mutual trust and one where the promises of the reasons we work can be fulfilled.

The technological advance that is fostered by those directly involved, along with its effect on work content discussed above, also has other very beneficial side effects. One of the basic forces driving technological advance is the objective of the reduction or elimination of dangerous and just plain drudgery type of work that must somehow get done. It is more than just work content (labor hours) reduction; it seriously upgrades the overall character of work. Let the robot or other machinery do the dangerous and dirty work. This requires an enthusiastic and mutual embrace of technological progress and its side effects by both management and labor.

In a "turned-on" organization where teams work together with all contributing to produce better ways of doing things, the atmosphere transforms drudgery to rewarding life's work. It may be hard for some to believe, but even on the production floor of an American auto plant in the 1990s, TGIM could be found. From its beginnings, management and organized labor agreed on principles that facilitated that kind of atmosphere at the G.M. Saturn plant in Spring Hill, Tennessee. It was unheard of in American auto companies.
Outcome:
- Best possible job security
- Character of work is upgraded
- Drudgery is minimized
- Work becomes truly rewarding

- TGIM becomes real (looking forward to going to work – can you imagine it?)

(It was hoped by many that the Saturn philosophy could spread to other G.M. divisions. Unfortunately, the promising early successes lost their meaning as the paradigms of both the UAW and G.M. came home to roost. It seems as though the typical auto plant mentality has spread to Saturn and for the most part it is now just another G.M. division.)

Organized labor in the U.S. has had a very significant impact on labor itself, but also on management's attitudes and philosophies. It has even had significant impact on non-organized work environments by association and by patterns developed over time from negotiated agreements. The natural resistance to do or contribute any more than has been negotiated in an adversarial environment is often present.

There can be no doubt that labor unions have been an absolute necessity. Workers were taken advantage of more times than not, even into the middle of the twentieth century. The workers were viewed as a distinct lower class and needed a champion.

Today, an atmosphere of openness and freedom of expression and devotion to life's work seems to exist in many modern companies. The promises of the reasons we work can be fulfilled and the culture of Yankee ingenuity prevails. That atmosphere is different from some in the so-called "rust belt industries," where the tension of fragile labor agreements and mindless drudgery can make eight hours seem like twelve.

"Work" in the modern industrialized countries has evolved to the point that protection to a large extent is built into the national systems. As a matter of fact, it appears that the reason for management to withhold more favorable treatment may be because of the defensive and contentious relationship that exists with organized labor. More favorable benefits may be traded for something management needs to be more competitive at the next collective bargaining session – certain work rules, for example.

Could it be that the very reason for the existence of organized labor has now become an obstacle for the realization of even better conditions? Consider that major non-union work forces in the U.S. auto industry are doing at least as well or better and have a spirit about their work that does not exist in comparable closed union shops.

Many modern companies have found that being able to harness the ideas and energy of motivated team members is mutually rewarding, and not just in financial terms. They are "turned-on" as opposed to "turned-off" in unenlightened organizations. Small to very large companies of all kinds, including automobile companies, operate very successfully in this way.

As this book was beginning, the newly elected President of the Teamsters Union, James Hoffa, vowed "militancy" to achieve union goals in his acceptance speech. Will militancy cause the many organizations involved to become "turned-on?" Will they be better able to compete, survive, thrive, be secure in their jobs and realize the many benefits of their life's work? Or is it a wedge to be pushed deeper between management and labor, in reality, endangering job security?

Labor unions fill the role of protector of their dues-paying members. The basic premise is that each member must meet the negotiated minimum job requirements. They then have the full protection of the mother union and gain the top objectives of seniority and retirement at the earliest possible time.

Yankee Ingenuity at the Organization Level

Let's take a moment and recap our observations up to this point: Yankee ingenuity has been instrumental in the American success story. Its "find a way" mentality results from the rewards of W3. Its primary ingredients are imagination, passion, and courage. It works because of the freedoms won for us in the American Revolution.

In 1822 John Quincy Adams said:
"Individual liberty is individual power" (empowers individuals) "and as the power of a community is a mass compounded of individual powers,

the nation which enjoys the most freedom must necessarily be, in proportion to its numbers the most powerful nation."

Adams neglects to tell us that the individual liberty that is ours is like a two-ingredient glue. It doesn't work unless both parts are utilized in the correct proportions. The second ingredient for realizing the benefits of individual liberty is the courage to exercise it and to pursue its advantages.

The "find a way" mentality or Yankee ingenuity at the level of the individual is pretty clear. Translating that to an entire organization is a different matter. The organization environment must be one where W3 is facilitated and fostered.

Seniority and Loyalty

In an organized labor environment, seniority establishes layoff sequence, regardless of contribution, as well as certain job preference advantages and retirement plan benefits. Otherwise, all are treated alike and employment is secure, regardless of individual accomplishment or performance as long as the minimums are met and the company survives.

Seniority filled with the fulfillment of the reasons we work is golden. Seniority filled with empty TGIF drudgery is a waste. It's like harvesting the grape skins and ignoring the sweet nutritious fruit inside or the wine or the raisins that could come later. The pursuit of seniority for the sake of seniority alone is a mistake. Some of us, after achieving some minor level of seniority, can feel secure and that we are on our way to retirement, regardless of the satisfaction level. It becomes a form of handcuffs and can blind us and prevent the pursuit of really meaningful objectives that will fulfill the promises of W3.

Thirty or forty years of mindless drudgery or fulfillment?

Each of us will place different values on the various reasons we work, probably unconsciously. Those individual values or priorities, therefore, will determine our goals in life – to be president, a green's keeper or a member of an auto-building team. The priorities may change as we mature as latent imagination, passion, and courage may be awakened by new experiences and accumulated wisdom.

The typical dues-paying union member surrenders decisions regarding his or her workplace's well-being to the union. In the process, many of the opportunities for the fulfillment of all the reasons we work are sacrificed. Most feelings of self-reliance, independence, and liberty are forfeited.

In modern business, loyalty is an equation. That is, both sides must be in balance or neither can exist for long

W3 achievement = enterprise mission succeeds
Enterprise mission succeeds = W3 achievement

In other words, management and the work force must each reciprocate loyalty, another benevolent cycle. It can be the electrification of the work place atmosphere where the spirit of W3 and the "find a way" mentality is pervasive. Some believe that lengthy seniority itself demonstrates loyalty. In fact, the two terms are unrelated. However, seniority is a natural and beneficial result of the equation balance.

Every enterprise strives for success; the alternative is failure for all involved. Success produces all the benefits to those involved: shareholders, customers, and employees. Enterprise success (the collective life's work) is fundamental to progress as defined in Chapter 2. Success is achieved by individual contribution at all levels. As stated earlier, the pursuit of W3 by its people is the most powerful attribute a company can have in its quest to succeed.

The labor unions in the U.S. have accomplished the mission that created them. The workforce in the Western industrialized countries is protected and better off than ever before, having reached their current benefits plateau. Labor unions played a major role in that accomplishment.

The membership level of that plateau, however, has had a downhill slope for several decades as membership has declined substantially during that time. The "turned-on" organization, where associates or team members participate fully and realize the fulfillment of life's work promises, is becoming a larger part of the total as the competitive landscape is transformed.

The challenge for organized labor and management today is to find the way to leverage their collective resources to provide for the fulfillment of the promises of the reasons we work for all involved. It is the next plateau and it will bear sweet fruit for all. Their executives', managers', and representatives' own chosen life's work should be directed to passionately facilitate that goal and the inevitable resulting job security and higher levels of company effectiveness. The benefits roll up from individual to enterprise to progress for humankind.

Job Security

Job security is a sacred goal of organized labor that negotiates for guarantees at every opportunity. What causes secure employment? It is obvious that the ongoing success of a company, in the increasingly competitive global business environment, is the cause of job security and prosperity for all involved. When a company fails, labor-organized or not, guarantees or not, the employees fail as well. Every company must have the best chance to succeed for the benefit of all concerned. Freethinking, contributing, spirited, passionate, and committed employees in an environment that fosters W3 create the best chance.

Company success is the only cause of job security for the collective organization. However, there is an even greater cause of job security for any individual. It is his or her personal reputation and record of achievement. Unfortunately, individual successes and failures are moderated and personal reputations are almost anonymous in many organized labor environments.

"The UAW, which has seen its ranks decline since peaking at 1.5 million in 1979" - "at the end of 2000 had 671,853 members"[14]

---

[14] *UAW ranks shrink in 2000* (Midland Daily News April 24, 2001)

That loss closely parallels the loss of market share by the American auto companies.

There are numerous causes for the share loss which are all reflected in those companies' ability to provide competitive value in their products which in turn is directly related to job security for all involved.

Some believe that job security is a right or a benefit. They seem to believe that it can be negotiated and that it requires little responsibility on their part beyond that which is specifically negotiated in their contract. They target world free trade and the low-labor cost countries as being unfair threats that apparently threaten American jobs and must be excluded or otherwise penalized for that perceived advantage.

The view from the other side, the less developed, low-labor cost countries, could easily be that the industrialized nations, the U.S. in particular, have an unfair advantage in technology and capital availability. Their advantages include their well-known capability to exploit emerging technology and create new and imaginative methods and hardware. Those characteristics provide an advantage over high-labor content. This is particularly true in view of today's rapidly advancing technology.

"A study by an international consulting firm found that the average labor productivity in the modern sectors in India is 15 percent of that in the United States. In other words, if you hired an average Indian worker and paid him one-fifth of what you paid an average American worker it would cost you more to get a given amount of work done in India than in the United States."[15]

Isn't the real answer that imaginative application of advanced technology *when teamed with imaginative, effective operation* (Yankee ingenuity) should in fact have an overall cost advantage on those low labor cost countries? While direct work content in the production of the product may be reduced, other jobs are created and many other benefits accrue to the overall workforce, not the least of which is company

---

[15] Thomas Sowell, *Economic illiteracy dangerous to third world*, (The Detroit News, July 21,2002)

success, even survival - preserving and creating secure and fulfilling jobs.

Chapter 5:

Historical Snapshot

The Impact of WWII

Before the start of the war, President Franklin D. Roosevelt called together a group of advisors to determine what would be required to launch the maximum effort to provide modern war material at a rate previously unknown. Newly-conceived weapons and ammunition of all kinds - tanks, vehicles, ships, and of course, fighter and bomber aircraft - would be required in vast volumes and on unprecedented schedules. These people were recognized manufacturing experts from those industries, with the auto industry being the focal point since mass production would be the key. The American auto industry was far and away the world leader in mass production, which had become a part of American culture. Included in an advisory capacity were prominent manufacturing engineers and special machine tool people who had played a major role in advancing mass production.

Nothing was sacred in this effort and every resource would be made available, including existing production facilities. They would be retooled and new facilities would be built. Many of the items to be produced weren't originally designed for mass production. They now had to be redesigned based on the advice of those who knew how to mass produce products. Components would now have to be made that would be interchangeable and fit any generic mating part that may come down an assembly line.

With the exception of the automotive industry, much of manufacturing at that time was still being done on a craft basis. Parts were custom-made to fit a particular mating part, which is not conducive to mass production. The American automotive industry had revolutionized manufacturing by adapting, perfecting and exploiting mass production techniques. The special machine tool is integral to mass production, as it is custom designed to produce precisely duplicate parts in high volume as first demonstrated by Eli Whitney.

Imaginative manufacturing processes were devised and the special machine tools to produce hundreds of thousands of different parts for these war goods were retooled or designed and built from scratch. There was massive subcontracting undertaken spreading the effort throughout the country.

Before Pearl Harbor and the actual declaration of war, the production effort was known as "Lend Lease." It supplied much of the material to the U.S. allies, including the Soviets. Joseph Stalin was said to have toasted American productivity as the key to winning the war. The effort was supercharged following Pearl Harbor and was unparalleled in history. Many, like Stalin, would say that it was responsible for winning the war.

A closer view of some of the specific accomplishments, new products, their manufacturing methods, and the time it took for them to arrive at the front lines for the first time would seem incredible, even today. There were no rules, precedents, or guidelines for satisfying these kinds of requirements. It was the miracle of mass production created by a culture of Yankee ingenuity.

The Boeing B-29 Superfortress was designed in 1940 and its maiden flight took place in 1942. It was by far the most advanced of its day. It could fly faster, farther, and higher than any other in the world. Production began in mid-1943 and by mid-1945, 3,765 had been built. Its builder was selected by a competitive bid and prototype construction process.

On three occasions, while on return flights from bombing runs over Japan, B-29s found it necessary to make an emergency landing in Vladivostoc in the former Soviet Union. The crews were eventually released, but the aircraft were confiscated. The Soviet air force trailed other industrialized nations dramatically, and they had to find a way to accelerate that capability. The Tupolev TU-4 was a reverse-engineered B-29, down to the one million rivets and including most avionics and other peripherals. It took two years to get to the maiden flight, about the same as the U.S. original concept. Boeing had been working on a concept privately before the war. The Soviets built nearly 1,000 over the next decade. Variations for commercial use were also built.

As often referenced, Admiral Yamamoto's worst fears would be realized – Pearl Harbor had awakened a sleeping giant.

"And that the enemy mind was once more becoming belatedly aware of its gross miscomprehension of the source of American strength became apparent on February 11, 1943," as Albert Speer took over as the German Minister of Munitions and Armaments - referencing American industrial might.[16]

What do you suppose happened to the rulebook during this period? It's certain that many traditions, standard methods, and paradigms were scrapped. The rules referred to are not those relating to integrity, morality, and fairness, but those relating to encouraging unrestricted free thought and finding better ways to accomplish their objectives. Yankee ingenuity involved more than just the engineering and technical aspects, but also the organizational aspects to find ways around traditional barriers that would restrict the extraordinary progress that would be required.

So now we have seen in three important periods in American history that when advances in manufacturing and technology accelerated to levels not previously thought to be possible, it was in a "no rules environment." The very beginning of industry in the U.S., the circumstances in the pioneering environment in the advent of the computer and related technologies and the demands of WWII had a common denominator: Imagination, passion, and courage; (Yankee ingenuity) unencumbered by

---

[16] Christy Borth, *Masters of Mass Production* (The Bobbs – Merrill Company, 1945) 84.

tradition, standards, rules, and paradigms produced innovation and dramatic advances.

There were no precedents to establish direction. In many other fields the early discoveries triggered similar circumstances, and new playing fields without rules came into being.

During the Vietnamese war, it became apparent that a way had to be found to patrol the 3000 miles of waterways that were the supply routes for the enemy. A very maneuverable, high-speed gunboat with a shallow draft would be required. Such a vessel was not known to exist at the time. It was felt that the pleasure craft industry might be able to come up with something close. A potential supplier meeting was organized and requirements given to the attendees.

One of those suppliers was building fiberglass pleasure craft similar in nature to the specifications except it was a different length. The boats being built utilized the recently-perfected water jet propulsion system that would be ideal in the shallow canals and rivers. The jet mechanism did not extend below the hull outline on an already shallow draft boat.

The company offered to build a prototype at its cost. The Navy asked for drawings and specifications of the proposed vessel before agreeing. The company's response was that they could have the prototype built and ready for initial testing in one week, but did not have time to make drawings, write specifications, or perform other bureaucratic requirements. The first gunboat prototype was in the water for tests a week later. The performance record of these boats built for that conflict was outstanding.

Do you think that in most circumstances that company would even be allowed to submit a proposal in today's business environment, either with the U.S. government or with business? Without the supplier's imagination, courage, and passion what would those boats have looked like, and when would they have been available? This is a classic example of what can result from open-minded people putting an important objective ahead of rulebooks, and paradigms - both the buyers and the seller.

Evolution

In earlier days of the automobile industry and other manufacturing industries, the responsibility for establishing manufacturing processes and specifying tools and machinery was, for the most part, retained by the producing company itself. Some industries today still prefer that vertical integration and continue their extensive manufacturing engineering efforts. Many of the tools used in those circumstances are adaptations or modifications of standard devices and standard machine tools that are readily available.

Driven by demand for higher and more efficient production, it became increasingly cost-effective to customize and even specify custom devices right from the beginning. There are numerous operations that could be performed only marginally on standard machines, but much more effectively on custom machines. There are machining operations that would restrict the product designer if standard machines were mandated and would severely limit the freedom to innovate.

The special machine tool companies in Europe (with Germany being the most dominant) and in the U.S., are, with rare exception, publicly owned, but are nearly all descended from smaller, family-founded and operated companies. The American engineer/manufacturer has a reputation for being imaginative and driven toward simplicity in devising the winning competitive solution. Their German counterparts have a reputation for designing beautifully complex mechanical devices, which can seem like the objective, rather than the desired, simple solution. They are also known for their high-quality craftsmanship. The Japanese have the reputation for improving on what they have seen, done, or purchased. In more recent history, their imaginations and resulting creativity have become an important factor in their success.

In Japan, the well-known machine tool companies are either spin-offs of the auto companies or are still a part of or very closely related to those companies. The globalization of business in recent years has lead to consolidations and foreign business ownership. The U.S., Germany, Italy, and Japan are the world leaders in special machine tools. France, Spain, and others contend on a smaller scale.

As is typical in family businesses, the managing owners were the experts in their field, as the knowledgeable and passionate vintner in a family wine business, and had a vested interest in the ongoing success. In many cases, these companies were initially "Tool and Die Companies." At first they provided simple tools, basically taking direction from their customers. At times, the customer even provided the engineering for the tool or die project. Many small tool companies still operate today, serving other manufacturing industries, including auto part suppliers, much like the larger ones did in their early history.

Until recent years, the American special machine tool industry has been the leader in supplying its products to customers worldwide. It has now been overtaken by the German and Italian machine tool industries, even in the U.S. market.

As the momentum built, specialty industries such as electric motor and controls, hydraulic systems, and of course, computer industries, recognized market opportunities and targeted product development to machine tool builders specific needs. Collaborative research efforts were undertaken with universities.

These small companies began the process of improving on what they were being asked to do to be more competitive. They could offer an alternative solution and get an order that otherwise would have been given to a lower bidder for the equipment as specified. Soon they became more expert than their customers. Their customers' efforts were diluted because other responsibilities consumed part of their time and efforts and their family fortunes were not a stake. They had different priorities.

Simple tools evolved into mechanized tools, increasing reliability, reducing the number of operators, space required, and cost, as reflected in the final product. These companies were fast on their feet in offering the next technological step forward on each competitive exercise to beat the competition and get an order. They were true entrepreneurs. Risk was always present and potentially deadly, since an order value could exceed the company's equity.

Lead by their manufacturing engineers, the automotive customers began to realize that important benefits could be gained in the free enterprise acquisition of special machine tools. The resulting competition accelerated the evolution of manufacturing technology and its hardware. The providers of the equipment supplied competing customers and other industries as well. Competition among suppliers, competition among users, and cross-pollination to unrelated industries and to those in the other industrialized countries leveraged technological advance universally.

In the earlier example of the chainsaw piston production machine, the buying organization requested quotations from special machine tool companies in various industrialized countries to compare costs. They asked for proposals for manufacturing processes and equipment to produce their pistons. In effect, these inquiries put to work the most experienced, competitive, and clever manufacturing engineers in head-to-head competition.

The result was a different approach from each of the respondents, one of which would be the successful bidder. A second builder of similar engines then benefited, as the successful machine builder now had greater specialized experience and product to sell. It could be an enhanced version of the first generation machine, incorporating improvements from lessons learned. Following that, the first buyer was able to get an even more advanced version his second time around.

The technology advance spiral is apparent. Certain unique features and characteristics developed on these machines would be applied to other projects. They would involve other customers in other industries and countries as they fell into place in the imagination of the machine tool entrepreneur.

The output of imagination in a specific area of endeavor is typically directed at solving a particular problem in isolation for that area - enterprise or industry. However, when that output (idea) becomes apparent to others, it will inspire its use or its variation's use in other areas for their own specific uses that were completely off the radar screen of the originator, even in other industries or other countries. The value of that first effort has begun to multiply. That value can be further compounded when two or more totally unrelated ideas are seen by another uninvolved originator in any industry in any country. That originator may then imagine concepts that foster the integration of those ideas or their variations for completely new outputs of human imagination.

The subject industry, an incubator of innovation in its own right, is in a unique position. It supplies imaginative solutions to numerous and different industries in many countries and against worthy international competitors. As such, it is part of a network that exposes it to those industries, countries and competitors and their diverse solutions to their diverse problems. That network exposure is an important resource that leverages local imagination, serves as a competitive advantage if utilized effectively, and is a driver of technological advance.

For the buyers of manufacturing process techniques and the hardware to implement them, the experience and imagination resource available to them in the special machine tool industry is limited only by their ability to evaluate the responses to their inquiries.

Those companies choosing to continue to do their own manufacturing, engineering, and special machine tool development and those that feel they are better able to dictate process and hardware concept to the machine builder will not benefit as much from this upward technology spiral. They could soon find themselves inbred, myopic, and stagnating in relation to their competitors as their primary resource is within their own walls.

The business relationship described does not typically involve the engineering and production of component parts of the buyer's products. Rather, this relationship provides the original, intellectual products of manufacturing and process expertise that will be manifested in their own hardware and software products. This relationship certainly cannot be unique to the automotive companies and their special machine tool sources. Most other product circles, such as semi-conductor businesses, chemical and plastics businesses, and even software businesses, will utilize the equivalent supporting kinds of businesses. The same benefits, not the least of which is the Yankee ingenuity "find a way" component that produces dynamic value, are certainly among their objectives.

Some customers' engineers are more reluctant than others to buy a new approach, due partly to "not invented here" pride and also due to risk aversion. The risk of something that they specified and purchased not functioning properly jeopardizes their job and reputation and promotes conservatism in selecting an approach. Continually finding and employing new and better ways does involve accepting some level of risk.

The Stress of Succession

Later generations of family entrepreneurial businesses were not always interested in or capable of succeeding the founding generation in managing and controlling business. In some cases, they were interested but did not necessarily have the natural gift or passion for the business or the intensity that it takes to succeed. The extreme demands of those unique businesses and the infectious nature of their challenges became a passion. The challenges drove the successful entrepreneur beyond the normal requirements for earning reasonable livelihood. It could be a really tough act to follow.

There are many examples of companies reaching that critical point in their history. That is, the first or maybe the second generation entrepreneur reached retirement age or incapacitation, and no capable heir was available to assume control. There are also examples where it was assumed that an heir would be ready and therefore, little serious grooming of non-family persons was undertaken. In addition, the heirs needed to find the way to cover impending tax obligations or get their inheritance capital out following or anticipating the incapacitation or passing of the entrepreneurial generation. Outright sale or going public were obvious alternatives.

The entrepreneur himself could be dominating and unwilling to relinquish much control or to share ownership. Aspiring young management prospects can become disillusioned and look elsewhere for opportunities, diluting efforts to build an enduring, visionary, world-class company. Those business environments were not necessarily the type that would encourage the pursuit of the W3 rewards.

In the case of special machine tool companies, many were left with control in the hands of professional managers from outside the industry. There are numerous examples of failures in these situations. The former manager understood the customer, the manufacturing engineering principles, and the competition. He was the passionate entrepreneur participating in the daily competitive battles, "The Machine Tool Guy," and the energy behind the company's success.

He also knew the competition and would try to anticipate competitors on each exercise, knowing their backlog (how hungry they were) and their technical strengths and weaknesses. Often, he could be autocratic to the point that most serious decisions were his and as a result, his people did not presume to interfere. The company may have been well-organized and contained supportive talent, but it had not been trained to make the top decisions or to envision and set the future course for the company. With the departure of the entrepreneur went much of the wisdom, spirit, and passion that were the competitive soul of the company and much of the industry.

A new manager from the outside was not capable of making industry-specific decisions on his own, and the organization did not have the depth or breadth to help, a formula for potentially serious problems.

Many of the closely-held companies could not manage the transition to the higher levels of business or the succession stresses. They found difficulty managing the capital requirements and going public due to the rust belt perception of prospective investors. Most companies were sold and the family heirs got their inheritance capital.

There are examples of other industries experiencing this same problem with more favorable outcomes. Their aspiring young workers could make their presence felt and gravitated to high levels of responsibility. They were able to overcome the obstacles faced in the special machine tool industry. In some cases, stock was made available to them over time and they became significant stakeholders perpetuating the entrepreneurial nature of the company.

In other cases, they saw that it could not happen where they were and opened their own businesses down the street, helping to perpetuate the industry. In fact, the natural market forces were enhanced by the added competition benefiting the buyers. There are examples where certain industries are centered in particular geographical locations resulting from these kinds of spin-offs. Furniture centers are an example.

In the Detroit area, there are a number of successful engineering companies, each having a nucleus of multiple owners/managers who left special machine tool companies in this way. There are a number of manufacturing support companies and machine shops, as well. The entry barriers in an engineering business consist largely of the personal reputations of the principal owners and managers and their contacts in their business communities.

The very high capital requirements of the special machine tool business are an entry barrier that is nearly impossible to scale for an individual. One such spin-off did happen in the late 1960s, and while the individual struggled initially, he was ultimately successful and his company grew substantially.

Not being directly consumer related, it is a very low profile industry. In effect, it is hidden from many imaginative and passionate young people, the Bill Gates or Mike Dells of the world, who could help to perpetuate it. In most cases, the financial barriers are simply too high without planning and assistance from the organization itself, as others have done, and so the talent and spirit have not gravitated to the ownership level.

The special machine tool experience typically involves working closely with engine builders and aircraft manufacturers, among others, from around the world. This is an exciting, challenging, and adventurous life's work environment. Can the disciplines involved in conceiving, engineering, and manufacturing special machine tools for diverse worldwide customers be less glamorous or challenging than writing computer code day after day? If the industry profile and its "glamour" had been at higher levels, would the Gates or Dell types been attracted and be managing world-dominating U.S. special machine tool companies today?

The Vitality Drain

The human body is made up of many elements. Many of those, while contributing to general health and well-being, are not essential to life. There are, however, microscopic elements involved that are essential to life. The American special machine tool industry is one of those trace elements in the body of our society. It is microscopic in its stature in the daily lives of our citizens, but its role is life-sustaining. An example is the role that manufacturing engineering played in the miracle of production for WWII materiel. What is to happen if there is another such national emergency? (At that time much of that capability still resided within the auto industry, but more recently shifted to special machine tool companies.)

Over time, a company or an industry can become less adventurous and more comfortable as "this is the way we do things" paradigms, standards, and traditions accumulate to the point where originality is stifled. In those companies that are mass producers that phenomenon is problematic, but in technology-originating industries, such as special machine tools, it will be lethal, since originality is at the core of every product.

Events such as WWII can set the paradigm register back to zero by disregarding traditions, paradigms, or rules. That fresh start reinvigorated American industry in general. Today it is not uncommon to hear the term "downsizing." It is a way for companies to set their paradigm register back toward zero and get that fresh start, if it is done correctly and not simply to take out excess cost. It is obviously less traumatic for all concerned if that freshening happens routinely and continuously by company philosophy and vigilance.

The challenge that all businesses face is lurking behind and is obscured by all the day-to-day challenges. It is the accumulation of standards, rules, traditions, and paradigms that encumber free thought, imagination, and the freedom to originate. The temptation to play it safe, add a rule, guideline, procedure, or department, or to let tradition dictate policy, methods, or procedures can be very hard to resist. These measures soon become repressive.

It is even harder to resist the customer when he advises or brow-beats supplying companies to be a part of or belong to an industry standard organization, but the same result can be expected. That advice and the spread of standardization also has the effect of raising nearly insurmountable entry barriers. It forces overbearing infrastructure on start-up companies and dilutes the desire of new, free-thinking players, the small, isolated populations, to enter a business.

The true value of an enterprise is in its success, as it benefits its customers, its owners, and its employees. The challenge is to be better than its competitors, not the same as its competitors. The freedom to be fast on its feet and to use its Yankee ingenuity in all that it does *and the way that it does it* makes an important difference. To insist that supplying companies, particularly technology-originating companies, fit a standard profile and other paradigms is counterproductive. The war materiel successes in WWII were as much a result of creative ways around organizational and administrative barriers, rules, traditions, and paradigms as they were of technological innovation. The creeping rigidity of the single huge population can paralyze free enterprise.

Yankee ingenuity thrives in the "no rules" environment! That environment must be conducive to and encourage free thought and the courage to confront challenge and risk and to endure the consequences of both success and failure.

Chapter 6:

Leaders: The Facilitators of W3

Business Management

Several things distinguish the special machine tool industry from others, factors that have an important impact on its management methods. The most important is that the typical order size can be quite large in relation to the size of the company accepting that order. It is normal that a company may only process a relatively small number of significant orders in a given year. There are always many small orders for spare parts and re-tooling components as well, but they are normally only a small portion of the total.

The markets that are served are very limited. The total number of significant customers, although very large and multi-divisional, dealt with in a five-year span could well be under two dozen, and of those, only a handful will dominate.

Add to that the fact that a large portion of new business is newly conceived, with inherent risks, to beat worldwide competition to win the order. It's obvious that every project must succeed, financially, technically, and in terms of timeliness, or grave and long lasting consequences can be expected.

As the entrepreneurial generations have passed and public ownership in the industry has become more common, its methods and success record have been affected. The professional managers have tried to make the cloth of the special machine tool industry fit the pattern of their business management education and background. The usual discreet financial indicators tell only a small part of the story. Success or failure of the enterprise depends entirely on the success of large and urgent projects the details of which are very technical in nature. Customer relationships are the first thing to go when a project is in trouble followed by diminishing backlog and margins. Remember the small total number of customers. Just one of those could be a third of the total market and it is certain that any serious trouble is telegraphed throughout the industry.

Recently, an executive of a venture capital firm was asked, "What do you look for in deciding if an organization is right for your commitment of capital investment?" The top two requirements that he listed were: The organization team must be obviously passionate about its products and its mission, and the top management and/or principle owners must be as knowledgeable about all aspects of the business as possible.

A theme of this book is the importance of the culture of Yankee ingenuity; that is, the application of imagination, passion, and courage and their roles in producing innovation. That should not be confused with creativity for the sake of creativity. Creativity resulting from that culture and a nurturing environment is a wonderful thing, but it must also pass the tests of viability - technically, financially, and in its timeliness.

The following are quotes from an historical publication of The Cross Company, a large and well-known American special machine tool company. The company has since been acquired by a large German machine tool company and is known as Cross Hüller.

"It is an industry that repeatedly defies textbook analysis. Its management skills are seldom interchangeable with other industries, and its mastery requires a unique mix of skills which cannot be acquired outside of its boundaries. How then does it attract men who keep it alive and make it grow and flourish? Simply, it appeals to that breed of man who finds adventure in its uncertainties and satisfaction in its non-monetary rewards. The kind of men who see mechanical design and machine building as art forms. Men who love things mechanical and the prospect of solving problems not yet defined with devices not yet conceived. The kind of men that possess the self confidence needed to do things that have never been done before even after repeated failure." "The reason lies in the word "special." It is an industry whose continued existence depends on a steady flow of new and workable ideas, most of which become altered by the process of debate on their way to becoming workable."[17]

While these quotes stress the importance of the human element, sadly, when the family entrepreneur stepped aside, a competent top management vacuum remained that resulted in the appointment of the top executive from outside the industry.

Continuous innovation is a prerequisite for survival in the special machine tool industry. Therefore, it must have a climate that facilitates and promotes it. In other industries where it is apparently not a prerequisite (look again), it may not get the attention that it could and maybe should. A significant difference, of course, is the market it serves; that is, the products are many times one of a kind and must succeed every time on their own merits.

It is apparent that the entrepreneurs who led the few surviving companies for many years achieved their real compensation in much the same way as Eli Whitney did. That is, the satisfaction realized upon achievement of very difficult challenges, one after another, that few others in global industry would attempt. An additional reward was the belief that the organizations that they endowed with that capability and had nurtured to success would survive them. Unfortunately, in many cases they have not. The ones that remain are in serious question, at the time of this writing.

"Recently, a Dutch psychologist tried to figure out what separated chess masters from grand chess masters. He subjected groups of each to a battery of tests: I.Q., memory, and spatial reasoning. He found no testing difference between them. The only difference: Grand masters simply

---

[17] *Of Mechanics and Machines,* (75th anniversary publication of The Cross Company June 27, 1973).

SWEET AND SOUR GRAPES

loved chess more. They had more passion and commitment to it. Passion may be the key to creativity."[18]

A graduate from a technical high school in the 1950s went to work as a draftsman in a privately-owned special machine tool company and soon found that many of the "reasons we work" were satisfied for him. He did not think of it in those terms, only that he was very happy and was learning a lot. He developed great pride in what he was able to do. He made a point of using all the space available in the drawing title block for his full name so that he would be identified with his work. Many used only initials, since a full name leaves little doubt about responsibility or ownership - good or bad.

He would spend his lunch hour eating a sandwich while walking through the manufacturing area, hoping to see the parts manufactured from the drawings he had made and talking with machinists and assemblers (tool makers) about how good or bad those parts were and if they worked as intended.

After some time as a designer, he was creating assembly drawings with the same pride and watching those assemblies being built and operated for a small part of a project. Eventually, as a project engineer, he accepted responsibility for executing the engineering for an entire project. Much of the time it was an all-new concept developed by proposal engineers against difficult competition to win the order.

It is difficult to describe the satisfaction that comes from watching the material arrive, then be assembled, piped and wired, and then to watch the new concept, device, or machine come to life for the first time and be readied for a long productive life for a customer. It may be a relatively simple $100,000 project, or it may be an extremely complex $60 million project. That birth is always a great source of pride for those involved.

Following the direct project involvement came actually developing the new concepts, based on experience and imagination. Winning orders for significant projects against clever global competitors by out-thinking

---

[18] *What happens in the brain of Einstein in the throes of creation?* (USA Weekend January 1 – 3, 1999).

them produces a special emotional high. The concepts offered utilized sound ideas based on a thorough knowledge of the organization's capability and were as rewarding as designing or building it. You can bet, however, that he was on hand when it was birthed.

This leader is a composite of some of those who became top executives of those American special machine tool companies. The sad part of this story is that an environment to mentor and foster passionate, capable people that could result in any real level of ownership and perpetuate the entrepreneurial nature of the business did not exist in these companies, a common problem of those times.

Those things that are rewarding to the designers, builders, technicians, and their leaders when a project comes to life and functions as planned also reward the owner when it is a private company in much the same way. It is the recognition of project success and is reflected in the financial results accordingly. Many times it also signifies that the risk of a forward technology step was the right thing to do and the company's future competitive position has been enhanced.

There is a parallel involving the heart surgeon whose payoff includes the satisfaction involved when a healthy patient walks away following recuperation from successful, pioneering heart surgery.

It will always be desirable to have a significant level of ownership by solid special machine tool people. Numerous examples show that professional managers with proven track records have not been able to perform for the public owners in the special machine tool business.

Sooner or later, pure family ownership will likely come to an end, as special machine tool companies grow larger. Family successors to the entrepreneur will not be available generation after generation. Heirs are then no longer involved in operating the company and will get their inheritance (after significant tax obligation) sooner or later. Large sums of operating capital will be required on an ongoing basis. Thus, public ownership, at least in part, is almost a must.

Succession Thoughts

A fatal weakness in American businesses, the special machine tool industry in particular, has been to staff companies in a way that provides for the perpetuation of natural passion, an essential ingredient for success. Strategic recruiting, coupled with challenging, encouraging, mentoring, and providing specialized training and education is a logical approach to solving this problem. Observing, identifying, and finding the way to challenge and accelerate the competitors among them in a way that is fair to all is important.

Most of all, it must be made clear that there is a way to the top levels and that appropriate ownership (a stake), that can be defined in various ways, is possible. The opportunities must be as good as or better than what might be available to the young people in other places, both in terms of competitors as well as in other industries. Competition applies to all aspects of business, including acquiring, shaping, and holding valuable key people, which will determine success or failure of the enterprise.

That was a lesson not learned or accepted by some of the otherwise solid special machine tool companies and they are vanishing as a result. Great opportunities and perpetuation of an important manifestation of Yankee ingenuity are going with them.

Is it possible that in the two larger companies that remain in the American special machine tool industry and with the many small companies doing business on a different scale, that a resurgence of the industry is possible? Will they learn from industry "lessons learned?"

As long as there are products to be manufactured for consumers, the need for the special machine tool industry will exist. As long as passionate and fertile minds can be attracted and rewarded to apply accelerating and exciting technological advances from supporting industries, the American industry can be revived and flourish and can provide gratifying life's work for all involved.

Recruiting

Wouldn't it be valuable to know what a prospective employee is passionate about? Or if he or she has even discovered yet what may ignite that fire of passion, or if it can even happen at all? Is it possible to shape this person to be that inspired musician? Discovering passion for life's work can awaken drive, intensity, and latent talent, can be more rewarding in ways than many can imagine and can provide an excellent livelihood at the same time.

"The greatest good you can do for another is not just to share your riches, but to reveal to him his own." Benjamin Disraeli

In a recent consultation regarding cardiovascular heart valve reconstruction, the surgeon described the procedure to a prospective patient. He did it in a way that made it apparent that the surgeon was at the top of his game and the top of his profession. He is known, admired, and respected around the world for his achievements. Despite his lofty position, his manner demonstrated that he had the fire in him and that regardless of his past accomplishments, he will continue to do bigger and better things for his patients and his profession. In the process, he instilled confidence in the prospective patient. He obviously experiences all of the rewards of his chosen life's work.

An aging naval fighter pilot was recently asked what will happen when his age and related health changes end his "Top Gun flying." He replied, "This (his naval life's work) is not a job or career, it's a life."

If you really pay attention, you can pick out the people with a passion for their life's work wherever you may look. You may see it in a wait person at a major department store or at your local restaurant, the janitor in your building, in a skilled trades person, a musician, or surgeon. It's fun to look for them. Most of them make it very obvious.

In a recent visit to a florist, two middle-aged employees finishing grave blankets and holiday floral arrangements confided in a customer that they loved this work that they had been doing all their working lives and hoped that they could continue as long as they were physically able.

They are rewarded by customer-expressed satisfaction and recognition, and are able to express themselves in their creations and otherwise find fulfillment of life's work's promises.

A visiting male nurse explained his route to his profession. He was formerly a fully licensed plumber/pipefitter and had advanced to a well-paying foreman status in a small town. A major power generation facility was under construction that promised long-term employment and attracted skilled workers from miles around. The project was unexpectedly canceled and most lost their jobs, including this man. He had a decision to make. He decided to go to nursing school. It took several years and much sacrifice, but he achieved his goal of becoming an RN. The labor market in his former occupation has since recovered, so now he could choose either occupation. The foreman's position would approximately double his income, but his love for the nursing profession and the feelings of accomplishment and fulfillment in helping those in need made it an easy decision to stay in nursing. In observing him work, his passion for his work is obvious.

When interviewing prospective employees for any job or position, how is it determined that the individual is right for the job, for the company in the long term, and for the prospective employee him or herself?

When starting operations in the Midwestern U.S., Japanese auto companies interviewed hundreds of employees to select just a few. What were their criteria? They have been very successful and have not had significant labor problems and the American employees are not represented by organized labor, despite numerous aggressive organizing attempts.

Competition is now more global and more aggressive than ever. How do we select the right people for the future of any company to help make it a winning and prospering organization and to provide them with rewarding employment in that difficult, highly-competitive environment?

Let's hire three people:

- Entry level mechanical technical person
- Sweeper
- Software engineer

Characteristics to be evaluated:
- Appearance
- Demeanor
- Training level
- Experience level
- Education level
- Intelligence
- Independent thinking capability
- Resourcefulness - imagination
- Passion - spirit

How should these characteristics be prioritized? Rearrange the list in order of importance as you may see it.

Would you believe that the reverse order for the characteristics is the right one for all three positions? It should have been stated that in this effort, we are hiring not just for the needs of today and tomorrow. We are providing for the development of future leaders and for building the powerhouse that we want for the future of the company.

It may seem that education is not valued as it should be. It certainly is valued and if characteristics such as passion are there, the chances are good that education is also. If not, education, general or specific, can easily be added while it may not be that easy for some of the other characteristics. However, there are numerous examples of highly educated people without the energy (spirit/passion) to use it effectively. There are also many examples of less educated people, in the beginning, gravitating to high levels of responsibility and fulfillment because energy and passion were present in their work.

There are examples of sweepers becoming top officers in special machine tool companies. Some of the most well-known founders and CEOs of giant modern high tech companies did not complete their college education. There are current examples of special machine tool executives completing their degree studies in their fifties and they are not alone. What was it that thrust these people forward in their apparently rewarding careers, if not education?

Education, training, experience, talent, and even intelligence are just tools that have little meaning without the human energy and the motivating force that will make things happen. They are our passion, imagination and the courage to exercise them.

Leadership

All of the above sounds fine, but a bigger problem will be created in the process that can be as bad as those solved if it is not done correctly. We may have created a herd of high-energy wild horses galloping in all directions, without consideration of the common direction or good. There must be a leader that gives direction and provides organization and control, someone whom the others must follow to survive.

Sometimes, the word team is used in business, which to some implies that all have equal voice and have a vote on direction. Imagine a professional hockey team managed that way. In professional football, each of the players on a winning team are passionate about their position and their role, but they are directed by a coordinating leader so that their efforts add up to a total greater than the sum of its parts. Players personalize their assignments based on their particular talents that the coach appreciates and that he can weave into the overall play. He will customize his playbook based on the actual talent and capability; that is, the expected contribution mix.

Can you imagine an orchestra with the most talented musicians available not paying attention to the conductor? The musicians have individual style that the conductor recognizes and plays on for the overall effect. Each member of these organizations is an important contributor and benefits from the collective success. They will catch that pass and score, or make that sack, or perform that solo in an individual effort as their passion, imagination, and aspirations guide them and they will be recognized for it.

The principal tympanist (percussionist) and soloist for a major city's symphony orchestra recently commented that "It's a great challenge to keep up with colleagues who bring so much imagination to their music."[19] Imagination with teamwork in the ranks of a symphony orchestra!

Ultimately, many may coach or conduct themselves. They then face the challenge of giving their players maximum freedom for creative input, but then making the tough decisions and still getting 100 percent effort from them for the project at hand.

It's very rare that an individual effort can ignore the efforts of others in the same organization without negative effects. In software development, for example, even though there may be a great deal of special custom code within a given task where the more individual imagination employed the better, at each end it must relate to the work of others and it must be integral to the fabric of the total. This is true of almost any effort.

The qualities – passionate, enthusiastic, resourceful, and independent thinking can be those of football players, orchestral musicians, or surgical team members who are a part of a larger success story and realize the rewards of the reasons we work.

A desired quality of leaders parallels the master chef who is able to combine many ingredients and in an almost magical way, can bring out all the subtle influences of meats, fish, spices, herbs, flavorings, vegetables, and fruits to produce a gourmet meal. In other words, a leader capitalizes on individual talents and passion in the team while involving each member.

The leader must be able to customize the effort based on the actual talents and spirit of the actual players available each time out to maximize the output. He must be able to supplement, replace, train, and mentor to fine tune or to add a touch of a spice to the recipe. No recipe works to the maximum possible extent right from the book without some

---

[19] Lawrence B. Johnson, *Tympanist drums up solo for DSO at Meadow Brook,* (The Detroit News July 7, 2001).

adjustments, based on the actual available ingredients, the specific unique circumstances, and the specific objective for each exercise.

Leadership (author's definition): Inspire, counsel, and guide behavior and activities; facilitate realization of "W3(why we work)" for all, while assuming performance responsibility for a group or enterprise.

Leadership:

- Facilitates "w3 – why we work" for all
- Facilitates and guides the culture of Yankee ingenuity in a virtual "no rules" environment
- Leads by example (demonstration of courage in doing the right thing every time)
- Continually tests by trial (challenge)
- Facilitates and requires ownership (a stake)
- Facilitates advancement as aspirations and performance indicate.

Is this another conundrum? As leaders, are our own jobs in jeopardy because we have done a great job? We have made a point of creating an organization of imaginative, courageous, and passionate people to whom we have said that a way to the top is there for you if that is your goal. Are we intimidated and challenged daily by ambitious, inquisitive people to the point that we feel threatened by them and question our own courage to do the right things? What is the right thing?

We will have already done the right thing as that set of circumstances has certainly given the organization the best chance for success. A common measure of a good leader is that he or she has positioned the candidates for succession to reach that capability level at an early time. That, of course, makes it possible for company growth or expansion and the current leader to move on to greater opportunities if that is the goal.

Opportunities become available as much from the expansion of the company as from attrition. This results from its success and is facilitated by its passionate and imaginative people. Rather than a threat, it is in fact energy that benefits all, including the leaders, and is another benevolent cycle!

Wisdom in a leader means sensibly accepting challenge and risk, learning from both successes and failures, guiding his team accordingly, graciously enduring the consequences of both and acknowledging the impact on others.

In May of 1998 Daimler-Benz acquired The Chrysler Corporation. In the years just prior to the acquisition, Chrysler was very successful in nearly all that it did. It was known for its innovative styling concepts and for its successful mini-van, sport, and retro style vehicles. Its innovative approach to supplier company relationships produced mutual cost reduction successes and technological progress. It seemed to have a natural Yankee ingenuity culture.

The track record following the acquisition was so seriously flawed that to some, including the new owners, it seemed that nearly everything had gone wrong. Analyses showed, in fact, that many different things went wrong. How was this possible in such a short time?

The secret to the earlier success, as with most, was a passionate leadership team carefully assembled over a period of time. That team disintegrated within a very short period following the acquisition. Some left by their own accord and some by invitation, but either way, the effect was the same.

A senior executive, well-known industry-wide as a "car guy" with an international reputation for his understanding of product and market, was one of the first to leave. It appeared to be by mutual agreement. His style of directness may have made it apparent to him that he couldn't be successful, considering the impending cultural shift. He was later sought out by the largest competitor for an executive assignment at the highest level, including the responsibility for product and market strategy, a move applauded by many in the industry and beyond. His mark was very apparent after a very short period of time in the new assignment. Another very high-level executive, the one responsible for the innovative supplier philosophy, was unceremoniously released not long after the acquisition.

The human element and its leadership are typically the reason for the success of any company, as was the case with Chrysler. An acquiring company will normally act to retain that leadership, potentially by contract, at least through a period where a thorough understanding of the organization's strengths and weaknesses can be gained. One business cycle makes a lot of sense, which in an automotive business, from product development to introduction, is several years.

Remember, these were known leaders with proven track records. In the example, each of the members of the departing top team were the key people in each of the important disciplines that directed the company, typical of the industry of which it is a part. They in turn had assembled their spirited organizations over some period of time. The individual organizations could have easily lost their way because of the loss of proven, spirited, and coordinated direction essential to their success.

Can you imagine each of the important disciplines in that company continuing their earlier successes after the loss of their proven and respected leaders over a short period of time? It seems unreasonable to believe that they could! Was it arrogance or simply a lack of understanding of the local culture that got in the way of reasonable logic?

As of this writing, this company is showing signs of a reasonable recovery and will likely be fine, at least short-term. The question of cultural adaptation remains. For example, will the average American consumer gravitate to the rear-wheel drive preference demonstrated by European consumers?

Collar Color

In some organizations, many "blue collar" and "white collar" people believe that these are two distinctly different groups of people and that they may not easily interchange. The shame of it is that there is absolutely no reason that people who are involved in manufacturing activities, in any assignment, through his or her own initiative, can't reach beyond into more responsible and exciting roles. There is an imaginary glass wall and ceiling between them that may be fostered by a "bargaining unit" people versus "non-bargaining unit" people mentality or a similar code of behavior and no doubt includes peer pressure. You can bet that the "suck up" hang-up is also involved.

That glass wall has a terribly negative effect on the people who may never achieve levels of job satisfaction that are possible or even realize that they are possible, and it denies their company the use of that latent capability. The wall suppresses the passion and imagination that can achieve greatness.

How many people working in the "blue collar" areas have the very desirable characteristics of passion and imagination, but it is not recognized either by themselves or by others? Fortunately, many companies do enjoy an atmosphere where the blue and white collar people are fully integrated and all realize the rewards of their life's work.

Working class (classic definition): The class of people who are employed for wages esp. in manual or industrial labor.

The very term working class and its classic definition can seem to imply negativity and a lower tier of society. It can imply that a level of spirit and effort less than the highest should be put forth, and that there is little connection in spirit to the outcome of that effort. Limitations for personal achievement are also implied.

Working class (author's definition): The class of people, who by their own choice and effort realize the promises of "Why we work (W3)" and as a result are integral to the process of progress.

Is there anything negative about this definition?

"The man who founded Sears did not come from an elite background. Neither did the founders of such rival stores as Montgomery Ward and J.C.Penney. Nor did Ford, John D. Rockefeller, Andrew Carnegie, John Jacob Astor or David Sarnoff. None of them went to college, and all of them began working as teenagers in lowly occupations, the kind our clever and smug intellectuals like to call dead end jobs. There are no dead end jobs. There are only dead end people – and most Americans do not fall into that category. America symbolizes, above all, freedom and opportunity for ordinary people."[20]

Were these not working class people? What about people like Gates, Jobs, Dell, and countless others? Don't they fit the definition also? Some started out in their garages or basement on a shoestring. Many risked everything, worked harder than most for years, and ultimately succeeded. The exceptions are some, but certainly not all, "old money" families and members of royalty, but they are rare by any measure. That status in no way assures any of them and, in fact, may be a handicap in achieving the great rewards of W3.

There is a different and more basic stratification of American and other democratic societies that better describes the tiers that do exist today: Those who strive to experience the promises of W3 and those who do not for whatever reasons.

The Success Machine:

Fuel = Reward - (The reasons we work – W3)
Energy = Passion – (refined from the fuel of reward)
Engine = Courage

> **Performance enhancing additives:**
> ⇐ Knowledge - education
> ⇐ Talent
> ⇐ Experience
> ⇐ Training

Guidance System = Imagination

It should be encouraging to any younger people reading this to realize that the primary ingredients in the "success machine" come from within. Add to that Thomas Sowell's thoughts on dead-end jobs and his comment that "most American millionaires did not inherit their wealth,

---

[20] Thomas Sowell, *Fourth of July Celebrates Freedom of Common Man,* (The Detroit News, July 4, 1999).

but created it themselves" and it becomes apparent that one's local environment or circumstances should have little to do with success.

Courage

There was one important item not listed above in the characteristics to be evaluated in recruiting new employees. It is as important as any on the list, but it is nearly impossible to see in an interview. It is courage, and it is an essential ingredient in what it takes to be a serious competitor, to lead other people, and even to seriously pursue W3 rewards.

In the beginning, courage is venturing opinions while knowing that criticism may result and then taking ownership (responsibility) and intelligent risks, knowing that failures are possible. Finally, committing company resources when failures are possible, knowing that the owners, the other employees, and their families will share in that risk and then getting back on the horse after a failure all demonstrate courage. If people are tested frequently, they will develop confidence as they learn and experience both success and failure and courage will follow.

A very solid and respected engineer has a proverb in large letters taped to a wall in his work area. It says something to the effect that we must be guided by recognition of our own limitations in what we do. When and how do we find out what limits us, and does failure establish limitation?

The following messages have appeared in the *Wall Street Journal* and in other publications and are appropriately included here under this topic.

"Aim so high you will never be bored. The greatest waste of our natural resources is the number of people who never achieve their potential. Get out of the slow lane. Shift into the fast lane. If you think you can't, you won't. If you think you can, there's a good chance you will. Even making the effort will make you feel like a new person. Reputations are made by searching for things that can't be done and doing them. Aim low: boring. Aim high: soaring."[21]

"Don't be afraid to fail. You've failed many times although you may not remember. You fell down the first time you tried to walk. You almost drowned the first time you tried to swim, didn't you? Did you hit the ball the first time you swung a bat? Heavy hitters, the ones who hit the most home runs, also strike out a lot. R.H. Macy failed seven times before his store in New York caught on. English novelist John Creasy got 753 rejection slips before he published 564 books. Babe Ruth struck out 1330 times, but he also hit 714 home runs. Don't worry about failure. Worry about the chances you miss when you don't even try."[22]

Characteristics such as passion, imagination, and courage, can compensate many times for physical or even mental shortcomings or handicaps. Unfortunately, the reverse can be true where an otherwise perfect specimen may lack those characteristics and, as a result, may not succeed. In some cases those attributes may not be easily awakened.

Communicating

While it may not seem exactly germane to the overall theme of this book, there is one negative factor associated with the high tech advances. E-mail, touted as a major benefit of the high tech era, can provide great benefits in speed and volume of communication and broadcast coverage. However, it can change the way the real message gets through and perhaps its intended meanings. In addition, it endows the composer with a false courage, as he or she may feel less inhibited by not having to look into the eyes of the recipient or not having to worry about an immediate negative response. It denies the originator important communication tools. It can produce unrealistic views of work, and life in general, without early objective feedback.

---

[21] United Technologies Corporation, *A Message,* (Reprints from The Wall Street Journal, 1981).

[22] United Technologies Corporation, *A Message,* (Reprints from The Wall Street Journal, 1981).

Remember the saying, "If you can't stand the answer don't ask the question?" With e-mail, the answer is at least delayed, and will likely come in a way that will not require face to face confrontation. Some will say that is good, since more ideas may be ventured. In addition, copies (cc's) are frequently broadcast to many who need not be involved, sometimes for personal agenda or political reasons, wasting the time of the recipients.

There is no substitute for face-to-face communication and watching and reading body language. Gestures, a smile, a frown, a nervous tick or facial expressions help to convey and emphasize meanings. Looking into the eyes or touching, like a pat on the arm or a handshake, help to create clear, unmistakable meanings and reinforce two-way sincerity. Instantaneous feedback is also an important benefit.

"Salem, 47, is an academician turned entertainer, a scholar in the field of nonverbal communications who says, 'Up to 80 percent of communication is not what is said.'"[23]

The most effective communication tools, the defining attributes of history's greatest leaders - voice qualities and charisma, are not available to e-mail communicators.

An atmosphere of trust and encouragement without fear of criticism results in ideas being ventured freely and through spontaneous debate, improved upon, all of which in turn will help develop real courage. Measured real time reaction to real time action and events is invaluable and an important component of agility. People can feel isolated or detached and not a real part of the team without physical or eye contact. There is an important role for e-mail, but it should be used carefully for routine communications and information distribution.

Voice mail falls into the same general category and the results can be the same. In addition, voice mail has led to an era of serious communication etiquette deficiency. Courtesy and professionalism have suffered seriously by the routine screening of calls and not being available when

---

[23] Martin F. Kohn, *Mental Stealth Psychologist-entertainer Marc   Salem brings his mind games to the Century Theatre*, (The Detroit Free Press, February 12, 2001).

the caller requires and deserves a timely response, positive or negative. It has led callers to anticipate a non-answer, and meaningless messages stack up. It has become a vicious cycle and too many times, important messages are not delivered timely, if at all. There are times when it is an absolute must that human contact be made, but this can be impossible with voice mail systems being utilized as they are.

Solid communications involve emotion and passion and need real time discussion, which are not possible with electronic mail. If it is essential that something be done and in a timely way from a communication, voice or e-mail are not the way as you cannot be sure that the message was received, believed or acted upon.

"Computers enable us to share more information with more people with greater speed. But the more we think of ourselves as "information processors" the more we may come to regard the mere act of communicating rapidly to be as important as the content of the messages.

As computer networks rise in priority, so does the risk that we will become "tools of our tools" a memorable phrase from Thoreau."[24]

[24] Steve Wilson, *Faster Computers Won't make Us Any Smarter,* (The Arizone Republic - as it appeared in The Midland Daily News, April 7, 1999).

Chapter 7:

The Business Enterprise

The Special Machine Tool Enterprise

What follows is an effort to elevate the discussion of the individual Yankee ingenuity characteristics of passion, imagination, courage and the "find a way" mentality to the enterprise ownership level. Ownership in this context is the assumption of responsibility for an organization including, but not limited to, actual ownership.

A fact of business life is that every "for profit" enterprise must consistently and competitively improve shareholder value to survive. In the case of a public company, the shareholder makeup is constantly changing, based on the Wall Street perception of shareholder value; that is, track record and prospects for the future. Most of the time, they are investors without personal or sentimental attachment to a company or to an industry.

Those of us who own stock directly are those investors or shareholders. Many more of us are shareholders as part of mutual funds, retirement plans, or those who have insurance policies whose assets are invested in stocks. So most of us, as owners of companies, expect not just security in those investments, our assets earned from our life's work, but a competitive return as well. Most would agree that is not an unreasonable expectation.

But shareholder satisfaction alone does not assure enterprise success. In fact, there are two other constituencies that are equally important. While the shareholders in one form or another exert the most direct, day-to-day influence on enterprise operations, the other two also apply very direct but different pressures on those operations. One of those influences is the customer, who must find competitive dynamic value in product or service offerings. The third influence, if you accept the philosophy of this writing so far, are the contemporary employees who must be able to satisfy the promises of "W3" competitively.

It's important to note that a common denominator for the three constituencies is the adjective "competitive." With all things considered, why should any of the three continue involvement in the enterprise with less? This is pretty basic stuff, right? Enterprise results must consistently and competitively satisfy all three.

That is a serious challenge for the cyclical and highly competitive special purpose machine tool industry and others like it. Add to that the industry's intense capital needs and its innovative nature and associated risks, and its investment attractiveness is further diminished. In the case of public companies, financial analysts see to it that all risks to solid and consistent returns are well-known. What matters is a track record of consistent success along with a favorable economic outlook.

Enterprise Personality

Every organization, whether religious, political, or business, has a "personality," planned or not. Many times, it is simply a reflection of the founding entrepreneur's personality or that of current management. It is really the responsibility of management to shape the company personality for success. It may be an energy-packed, "find a way" environment with a "bring on the competitors" attitude. It may be a TGIF or a TGIM kind of organization. Even sub-divisions of larger organizations have their own unique personalities, which may or may not coincide exactly with the larger entity. The company personality is reflected in the way the particular entity functions and, of course, in its effectiveness in achieving its objectives. It also leaves an impression on prospective customers, prospective employees and prospective investors.

Many companies have a prominently displayed mission statement and maybe a companion policy statement of their intended integrity or values profile. While a lot of thought and effort may have gone into their choices, they may not be an accurate reflection of the real life personality of the organization. It may be just wishful thinking without thoughtful investment in the human capital and an actual operating philosophy that supports the statements that will make it truly effective.

So what does a desirable organization personality look like, and how can it be developed? The venture capitalist got it right when he required management knowledge and expertise and employees to be passionate about the company mission and its products before committing capital investment.

The statements of mission and values are an important start. They must be realistic, concise and something with which people will identify. The basic elements of success should be embodied in those statements.

Enterprise success (author's definition):
- **Price, quality and technologically competitive products - dynamic value (for customers)**
- **Competitive returns (for shareholders)**
- **Competitive "Why We Work (W3)" rewards (for employees)**

The mission and values statements are the rumble strips at the edges of the highway that will help keep us awake and headed for the enterprise objectives. The speed at which we travel toward those objectives is determined by the energy levels generated internally. The ingredients of Yankee ingenuity are implicit in the elements of success. However, they should be explicitly mentioned as well, as they will generate and inject that energy into the enterprise personality. It may be difficult to be concise with all these things to include. However, the concepts of continuous refreshing of organizational spirit and minimal rules, traditions, regulations, and standards, things that keep an organization energetic, vibrant and youthful (no reference to individual age), should be included.

If the organization's mission statement contains those points and is able to individually achieve them, then by definition, it has succeeded.

A large part of business public ownership belongs to institutional investors. They are mutual funds, pension funds, etc. who invest unemotionally using statistics, forecasts, and trends and require consistent and competitive returns. Even the largest special machine tool companies are probably not large enough on their own merits to interest institutional investors. Insiders and smaller investors are the most likely public owners of companies like special machine tool companies. They will also require competitive returns although some cyclicality may be tolerated.

In the case of a private company, which typically would be owned by someone who understands the business and all the positives and negatives, some patience with results is possible, but sooner or later there must also be payback. In rare cases, a poor financial performance can be allowed to extend for some time, depending on the depth of the sentimental attachment and commitment and the expected turnaround timing.

During the good years for the industry, when most special machine tool companies were closely held, they did well and were profitable most of the time. There were some good years and some not so good years, but on average it was a good business to be in for the typical owner, but probably a less attractive one for the typical Wall Street investor.

One of the largest five largest special purpose machine tool companies during the 1970s took an order for a system to produce disc brake rotors for one of the Big Three auto companies. The machine for the finishing operation was proposed using a whole new approach. There were two traditional competitors who were both well-liked by the customer and offered a more traditional approach. This new approach made the entire project more competitive than the competitor's offerings. In the eyes of the customer, the risk was whether or not just one machine, for the finishing operation, could accomplish the required production volume. In the view of the owner/entrepreneur, it wasn't a severe risk. He guaranteed that if it did not meet the requirement, a second machine would be provided at no cost to the customer. The system performed well at the required production rate with one machine.

That same company won an order for piston production equipment from a well-known small engine builder. There were two machines involved, valued at several million dollars. A competitor acquired this company as this order was entering its final assembly phase. The machines languished in the assembly and tryout stages. The acquiring company's president at the time did not have a machine tool background and wasn't sensitive to the fact that piston equipment was a specialty of that acquired company. When choosing people to be retained following the acquisition, he missed the specialty competence and focus that would be required.

As it turned out, another factor, perhaps too subtle for an inexperienced manager to see, had a serious negative effect on the project. The equipment was purchased for upgrading production and quality of a current set of piston products. With proper focus from knowledgeable people from the beginning, the equipment could have met the letter of the piston product tolerance requirements and probably met the delivery date requirements as well. As it was, some of the specific capability needs, statistically, fell short and time ran out.

This kind of thing can happen, but if the need is there on the part of the customer and knowledgeable people are involved at the management levels of the supplying company, compromises can normally be reached. The difference here was that the customer had the capability to produce those pistons on current equipment. It seemed that his production demands had tailed off and his capital could be better utilized on other projects. The project was very late and unable to meet the letter of the purchase order requirements, and he canceled the order. It was a disaster for that fiscal year.

This was a failure due to project management and lack of focus, not a technology problem. Had the rotor project not had the focus or attention, it would have failed as well, since there was a technology stretch involved. The owner knew his organization could get the job done if properly focused, and he would see that it was.

111

A pure public company would have some difficulty with both of these examples. An outsider's view of the risks associated with the level of continuous innovation makes bankers and investors very nervous. The philosophy of offering advanced competitive products to worldwide customers at some level of risk differs greatly from having to impress distant financial analysts monthly. The objective is to win orders that may be challenging but with the knowledge that it will lead to opportunities to profit from the experience

It is also a very capital intense business, requiring capital to fund the actual order production, potentially in the range of millions of dollars. It also turns over at a slow rate due to equipment deliveries that may be 12 to 18 months or more. The facilities required are very capital intense as well.

Historically, the owner/entrepreneur had a special rapport with several levels of management in key customer organizations. That was important to both sides, as both had very high stakes in the outcome of these projects. Those relationships continued over many years and projects and were proper in every way, the result of mutual respect and mutual support. They followed subsequent generations of management on both sides, and frequently resolved otherwise impossible conflicts. If such a relationship had existed and communications were working, the piston project probably would not have failed, even if it had fallen short of the letter of the requirements.

In a recent business year, a family-owned American special machine tool company announced to its employees that it would break records for that year for business volume and for profits on that business. That same announcement said that all those profits would be reinvested in the company for engineering and manufacturing facilities, which was not unusual for that company. The facilities prior to that announcement were already among the finest in the business. That same company has provided the most advanced special machine tool products of any competing company anywhere in the world.

The success of this company would likely be measured differently from various perspectives. The announced profits would seem modest to many; however, the shareholder value improvement would be significant, whether public or private. The net worth of the company was improved, its sales volume grew, and it became well-positioned in the global marketplace for the sale of its products. It was not unlike some public companies that do not show bottom-line profits year after year, but at the same time shareholder value increased competitively by growth and expansion through acquisition and its resulting market positioning.

A special machine tool company as a division of a larger multi-industry corporation can succeed under a management philosophy similar to investors favoring diverse portfolios. That is to say, the good and not so good years will still yield favorable results over time and provide the advantage of diverse corporate holdings for overall stability.

Corporations that demand specific levels of profitability year after year from all divisions, under threat of spin-off or organization-ravaging cost cutting measures, will not be a good parent. There will always be good and not so good years in the special machine tool business. There are times when two or more new automobile product programs occur simultaneously, and times when few significant programs are active. In the lean years, orders may be pursued which will require a technology leap to counter aggressive competitors. There will be difficult decisions on how to endure a low or no profit year, and to keep the organization together. That's very hard to accept in many demanding corporate cultures, but it is a driver of progress.

Competing

Any company's success is in its ability to win against its competitors. It becomes more and more of a challenge as the global economy becomes more of a factor. Alternative methods being developed by very competitive domestic and foreign suppliers in the auto industry, for example, challenge the traditional ways. Third World countries will be very difficult competitors for a long time to come. Do you remember those 50,000 indentured servants in England who made nails in the traditional way prior to an early Yankee? The thirteen American states were the equivalent of a Third World country of today. The free global market makes the rules, as it should, and will drive progress for a long time to come, whether we like it or not.

In any competitive exercise, the buyer or the decision-makers normally have several choices. There is one winner and all the rest are losers.

We all enjoy the fruits of competition in continually improving high dynamic value products and services that we need and buy every day. From a selection, we buy those of superior value from company's and their employee's efforts to find the better ways to win our business time after time. From airline equipment and services to breakfast cereal, whether it's an American product or one from another country, we will and should buy the best for the least every time (the best value). It drives the perpetual improvement process.

The "buy American" slogan and sentiment is ill-advised, and actually hurts job security as well as product value. If we continue to buy products that are not the best value, what incentive do those companies have to get better? In buying out of some sense of loyalty rather than for value, we subsidize inefficiency, give incentive to produce substandard goods and services and encourage rather than treat the illness. It becomes a handicap to those companies and is a serious disservice to them. Eventually, it will come crashing down on them as the real competitors will continue to get better and better and the buyers will switch one by one.

Imagine what a 2003 model American car would be like if foreign companies had not been allowed into the U.S. market. Consider their quality, driver and passenger conveniences, technology, and styling. Would we still have poor body fits, rust after the first year, and breakdowns? What would we have to pay for those 2003 models without the aggressive foreign competitors? The foreign companies weren't responsible for those advances. It was us, the discriminating, value-oriented buyers. Their early models had problems not unlike the domestic brands, but their determination to succeed in this market dictated a strategy of continuous improvement forcing response from local competition.

114

If all the countries of the world had a "Buy locally policy," the great American companies like Caterpillar, Coca Cola, McDonalds, Microsoft, Boeing, and countless others would not exist as we know them. Some of those companies sell more outside the U.S. than inside. They compete head-to-head with excellent foreign companies on their own turf and are better companies because of it, benefiting all. In addition, the competition, including foreign competition, that drives new product development and value to the consumer would be seriously damaged. The American automobile quality and value upward spiral in recent years is due to competition providing better value, and is just one obvious example. Buying the best value every time should be the top reason for the selection. That choice actually helps all companies equally who are paying attention and responding to the market messages.

Buyers of special machine tools also benefit by virtue of the competitive bid process and realize the best solution for their manufacturing process from competing expert companies around the world. Every order that is bid for is very important and the competitive pressures intense. There is only one winner each time. Any nationalistic influence, as in consumer product purchases, can only dilute the value to the buyer. As purchases are made on the basis of best value, a strong message is being sent to those companies that did not get the business. The message is that the product value was not competitive. The natural conclusion is that a better way must be found so that next time the competitive position will be favorable.

Ownership Examples

Two of the larger American special machine tool companies remain at this time. One is part of a public corporation, and the other is privately held. They are both vulnerable to the causes of failure being discussed. The privately held company must be able to manage large volumes of business and large projects from a working capital standpoint. In addition, there must be continuing progress in management succession following the passing of the family entrepreneur. Both are difficult challenges.

The public company, while diverse to some extent, has a great deal riding on their special machine tool and related businesses. So it must succeed, as financial analysts and public owners are demanding bosses and not particularly sensitive to the business itself or its human side. The top management does not have the depth of special machine tool background that has been traditional in that company or in the industry. There is considerable experience and talent remaining in the ranks. Are they experienced decision-makers or did former management make the important decisions? Their depth and breadth and just how well they are utilized and managed remain to be seen.

At this time, the global marketplace for special machine tools is in deep recession. At the same time, the American dollar is still very strong, especially against the yen and the Euro, making exchange rates a nearly impossible challenge. Survival for these companies is in serious question.

Ownership Model

Based on the preceding discussion an ownership model for succeeding in the global special machine tool market follows:

- Ownership, to the extent of control, is in the hands of employees, largely the managers at senior levels.
- The board of directors is composed of inside and outside directors, ideally including a majority with direct experience in the industry and some with background in finance/banking.
- The Chairman is a retired high-level special machine tool executive.
- The CEO with the characteristics described in Chapter IV is a seasoned veteran of the company, with an obviously solid track record, gravitating to top levels on his own performance. He is an enthusiastic proponent and facilitator of the Yankee ingenuity culture.
- A significant portion of manager's income is an incentive component and paid in company stock.
- Company match 401K paid in company stock.

- A stock option plan is in place for executives and certain managers.
- Stock purchase plan is in place for all employees.
- Highly selective hiring principles and training, mentoring and challenging are company hallmarks.
- Policies for advancement and ownership are stated in company mission, as is the Yankee ingenuity operating philosophy.
- A significant level of stock is held publicly for capitalization purposes.

Notes: Broad employee stock ownership is intended as a method to achieve ownership as much in spirit as in finance and should not in any way limit the stockholders from making intelligent investment decisions based on their own circumstances. Timely internal communications regarding company performance and prospects are essential. As stated earlier, the "no rules environment" philosophy cannot affect the integrity or financial responsibility and reporting or forthright employee, shareholder, or customer relationships.

This model of a company, which ensures management succession through "the reasons we work" recruiting and management, leadership trait fostering, and ownership, enabling and facilitating the "find a way" Yankee ingenuity philosophy, is the winning model.

This company is operated by owners who are knowledgeable about the industry and who are significant and passionate stakeholders. All the special things done by the earlier entrepreneur, from instilling confidence in the high level customers to visionary planning, to assessing the viability of and managing execution of projects and celebrating successes can be done by the operating owners.

Many of the employees are financial stakeholders as well as in spirit, since ownership occurs at all levels. The understanding and acceptance of the responsibilities to customers, shareholders and employees must be pervasive. In an ideal world, the family entrepreneur is retained on the board of directors and/or as an operating consultant through a transition period to broaden and harden the new managing executives and may retain stock.

The venture capitalist's top two requirements bear repeating at this juncture: the organization team must be obviously passionate about its products and its mission, and the top management and owners must be as knowledgeable about all aspects of the business as anyone can be.

This is hardly an original plan, but it is one that could have saved some of those that have disappeared. It should be considered by some of those companies in the upcoming cycle of smaller companies that may have succession concerns.

Chapter 8:

Buyers and Users

The Products Produced

Automotive Industry	Heavy Vehicle Industry	Miscellaneous Industries
Cylinder blocks	Cylinder blocks	Railroad wheels
Cylinder heads	Cylinder heads	Railroad axles
Connecting rods	Connecting rods	Locomotive cylinder blocks
Pistons	Pistons	Locomotive cylinder liners
Bearing caps	Bearing caps	Electric motor frames
Front covers	Front covers	Bicycle crankshafts
Intake manifolds	Intake manifolds	Power turbine components
Exhaust manifolds	Exhaust manifolds	Power generator components
Crankshafts	Crankshafts	Paper mill rolls
Camshafts	Camshafts	Aircraft wing spars
Transmission cases	Transmission cases	Landing gear components
Differential cases	Differential cases	Wing riveted fabrications
Valve bodies	Valve bodies	Wing and tail fiber layup
Universal yokes	Front wheel spindles	Small engine cylinder blocks
Universal crosses	Front wheel knuckles	Small engine cylinder heads
Front wheel spindles	Front axle I beams	Small engine crankshafts
Front wheel knuckles		Small engine pistons
Upper control arms		Small engine connecting rods
Lower control arms		Air compressor pistons
Rear axles		Air compressor housings
Rear axle tubes		Hydraulic pump bodies
Ring gears		Hydraulic valve bodies
Pinion gears		Uranium bars
Steering gear housings		Petroleum fittings

Steering gear rack nuts	Rock bits
Starter housings	Typewriter frames
Alternator housings	Military tank hulls
Air pump housings	Projectiles
Brake drums	
Disc brake rotors	
Disc brake calipers	
ABS valve bodies	

These lists demonstrate the basic industries around the world that are served by the special machine tool industry and some of the products that are produced on those machine tools.

The list relates to the actual manufacture of components. It does not cover the automated mechanical assembly or the welding of components to become larger, more comprehensive components. Nor does it include automatic testing, which is many times part of a manufacturing process. Together, they will likely equal the size of the list above.

It is not easy to put this into perspective, but consider the following: In the U.S. alone, in a 15 million-unit car year, the automotive list above represents roughly 750 million complex, precision parts to be manufactured, mostly on special machine tools, in one year. The small engine industry, consisting primarily of lawn mowers, snow blowers, chainsaws, pump and generator sets, etc, would account for a number of a similar magnitude.

Further, most of the components listed require multiple operations, usually meaning multiple machine tools. Many times, as in the case of a cylinder block for an auto engine, they will require many. Fourteen different multi-station machines are normal for the cylinder block, each of which performs many different operations. Similar numbers are required to process a crankshaft. It is also common that two or more lines of production equipment can be required to meet the desired production volume of a given product, doubling those numbers or more.

The components being manufactured are very diverse and range from delicate aluminum pieces, such as pistons and typewriter frames at very high production rates, to gigantic cast iron diesel or welded locomotive cylinder blocks and power generation turbine components at very low production rates.

The Automotive Customer

The automobile industry is easily the largest influence on the special machine tool industry. This is due to the fact that the competition in that industry drives frequent model change and, of course, because of its sheer volume. In addition, there are outside influences that necessitate frequent power train (engine, transmission, differential combination) improvements. The power train changes reflect directly on steering, suspension and braking and in turn create changes in them as well.

Courtesy Ingersoll CM Systems Midland, Michigan

Oil prices, fuel economy and availability, and government emission, fuel economy, and safety mandates dictate design changes. In addition, technology advances have given power train designers opportunities to incorporate new ideas. For example, the advent of computer engine management and diagnostics has changed certain mechanical aspects of modern engines, not unlike the computer impact on modern machine tools. As a result, the automobile companies have had to change complete manufacturing systems, not because they were worn out, but because their product has changed and different equipment is required. That makes it apparent that flexibility, versatility, or agility would be nice to have in a manufacturing system. There are significantly different levels of flexibility and different ways to achieve that objective, which are discussed in the addenda.

Each round of change for an auto company presents an opportunity to raise the bar for engine feature and performance competition; that is, market appeal. It is also a great opportunity to take the next step forward in manufacturing technology and cost improvements through the competitive bid process with the special machine tool suppliers. It is an upward spiral for both.

Other manufacturing industries operate on different competitive cycles, do not face the annual or frequent consumer-driven model changes and are not affected by government mandated issues such as emission or fleet average requirements. So the special machine tool industry is really dominated by the automobile industry. In recent history, there is typically a program or programs underway occupying capacity for machine tools. There are times when programs from different auto companies are in phase and times when they are out of phase. In other words, there are feast and famine times.

Serious strain on the machine tool companies has come with both extremes. Since it is a highly technical and specialized business, reducing staff in a slow market means reducing capacity for the future, since experienced, capable staff is its most valuable resource and is very difficult to replace or rebuild. So imaginative ways are required to hibernate in the winter of business cycles, losing only fat and not muscle. In the opposite extreme, long hours and heavy, well-managed sub-contracting can be required to avoid an organization build up that could not be managed during the slow cycle.

The heavy vehicle or diesel engine industry has some of the same influencing factors. Due to the lower volumes and lower direct dependence on the consuming general public, their change requirements are substantially lower. Their new engine product programs are maybe a third or a quarter as frequent as automobile engine programs. It may be coincidental; however, it seems that they wait for the automobile low purchasing activity phase. Then they find bargain prices, which is good for them and good for the special machine tool industry. The machine tool companies cover their costs while keeping their organization together.

An observation that seems to be true in the diesel engine business and some others gives them some advantages. They realize lower costs because of their ability to buy at the right time, and because the auto companies have processed several more programs between them since their last one. They get but do not have to pay as much for the technological advance developments that have occurred while their manufacturing engineers just watched and learned.

Still other manufacturing industries use equipment several generations old, because it continues to function reasonably well. Their paradigm of "if it ain't broke, don't fix it" prevents them from taking advantage of developments subsidized by the auto companies and the free manufacturing engineering possible through competitive quoting. Remember the chainsaw piston example: a tremendous payback simply by asking the questions and energizing clever manufacturing engineers from around the world who have likely advanced considerably since the last contact.

Since the automobile companies do indeed dominate the special machine tool companies, we need to take a close look at that relationship. It has changed substantially in recent years. There are many influencing factors. Changes in both the buying and selling organizations have negatively influenced relationships.

Procurement Practices

Because of the consolidations that have occurred in the special machine tool industry in recent years and the subsequent passing or retirement of the industry's executive entrepreneurs, customer relationships are less friendly. Customers at all levels have not felt at ease when discussing their production problems with executives in the industry who are not as knowledgeable as their predecessors were. The mutual respect and mutual assistance relationships that existed are now rare and relationships can even be adversarial.

The process for obtaining quotations and selecting a supplier for new equipment, while never perfect, was for the most part professional and honorable. The sanctity of the sealed bid principles of confidential price and the solution offered was respected. The special machine tool industry lives or dies on its ability to outsmart worldwide competition to win business based on price and concept. The sealed bid principle is fundamental to its success, even to its survival, and history demonstrates that it is the motivating force that drives progress – the dynamic value of the competitive offering.

Today, the industry is subjected to auction tactics and inappropriate use of intellectual property. That sanctity is a basic tenet of a modern free enterprise system. Ironically, that system was a very potent force in advancing the buyer's own interest for cost and technological effectiveness, but it is apparently not understood, appreciated, or respected as it once was.

All this seems very obvious, so why do some of the auto companies persist in these methods? Apparently, they believe that the effect is limited to getting the lowest price at the time. They don't seem to accept the fact that it pollutes the integrity of their procurement process. It also does serious harm to an indispensable supplier industry, seriously aggravates the already adversarial relationship and above all, seriously dilutes the value of their purchase. It is the "low price paradigm." It blinds the buyer to true dynamic value and is an impediment to progress.

Some in automobile purchasing management categorize these purchases the same as the purchases of automobile components in high volume from production suppliers in various and low labor cost countries, not appreciating the difference. The invaluable benefits associated with special machine tool acquisition done in a professional free enterprise way seem invisible to some of them.

The real products of these international special machine tool companies are intellectual in nature; they are the winning imaginative solutions. The associated hardware and software is the manifestation of those ideas.

The auto industry has been through alternating phases where in one, the engineers have the selection responsibility and the next, the purchasing management has it. Invariably, a buying company recovering from difficult times or in a cost cutting mode, as most are today, gives their purchasing department the cost reduction mandate in all purchases.

The offerings in response to equipment inquiries come from competing companies from around the world with content and currency exchange rates varying dramatically, thus benefiting the buyer. They can have entirely different concepts to accomplish the desired result. Two cost components are involved. The first is the total cost of the equipment being purchased, and the second is the cost of its operation over the life cycle of the parts to be produced. Together they are the real total cost. Many factors, mostly technical, including overall viability, are involved in that evaluation. The analysis should be made by manufacturing engineers who have a good grasp of those factors and with support and concurrence from the purchasing department.

The Auction

After submission of fixed price bids, a target price is established, lower than the lowest original bid, an arbitrary number. All acceptable bidders are then asked to meet or beat that price. Several rounds ensue, each at lower prices, until only one bidder remains. Final purchase order values can be a fraction of the successful bidder's original quotation, which has become a meaningless exercise. The best technical offerings often are not a factor. The other factors in the offering that affect the life cycle cost of the product to be produced may not be factored into the evaluation. The bidders are seldom invited to explain, present, or discuss their offerings to those involved in the evaluation. The written bid on a very large and complex proposal cannot possibly cover all the fine points of the offer, particularly if it's an imaginative fresh concept.

Imagine comparing bids that utilize dramatically different but effective approaches and then shaking down the bidders in an auction for the lowest price. One may obviously be better in the eyes of the engineer. He or she may not be willing to recommend the higher initial price, even though detailed study may show that the operation cost component offsets and even improves the actual total cost. Cost-sensitive management can be preoccupied with the very visible initial price readily compared to a project budget. The complex extrapolated total cost over the component life cycle can be a challenge.

The purchase price alone was more important than all the other factors including any thoughts of exploiting the culture of Yankee ingenuity for their own benefit.

Payment Terms

The trend in some manufacturing industries has been to purchase entire process capability at one time, rather than as single process steps as it was done in the past. In recent history, entire machining systems costing $60 million or more and even multiple systems of that size are being purchased from one supplier to focus responsibility and to minimize coordination problems and staffing needs for the buyer. The special machine tool company's ability to manage the capital requirements over the life of large projects has become a very serious problem.

Close relationships with banking institutions are essential, and the condition of progress payments from the buying company may also be required. Closely held companies are at a distinct disadvantage in this situation, as larger public companies typically have a greater resource base to manage this problem. In the case of foreign competitors, their bankers, customers, and governments have historically been more supportive than their U.S. counterparts.

The alternatives for private companies are to decline to quote the large orders, to consider going public, or to sell out to a larger company. Not participating in the larger projects sends a message to clients and to competitors that will lead to decline as the competing companies will grow and prosper and gain valuable experience and reputation. The smaller companies surrender as a major player. Is this what the buyers, beneficiaries of those imaginative technological advances, need or want?

All of the U.S. automobile companies discourage the progress payment option, and most absolutely refuse to consider them. If you accept the basic premise in this writing of the value of the subject industry to the automotive industry, the question of progress payments is a no-brainer. The selling price in a progress payment deal is always discounted for the cost of money. The pay back is invaluable. Other industries operate on the progress payment principle very effectively for the same reasons. The average homebuilder typically operates with progress payments, and you can be certain that builders of high-rise office buildings, bridges, and hydroelectric dams do as well.

Progress payments can substantially lower the special machine tool industry entry barriers, reducing operating capital requirements to help make possible new ownership blood.

Eli Whitney's musket enterprise and the related development of the first special machine tools would not have been possible without both up-front and progress payments from the U.S. government. Who can say what delay in manufacturing progress and our "more abundant lives" might have resulted by denial of those payments.

Consolidation and Relationships

One of the five or six largest special machine tool companies in the 1960s and 1970s specialized in piston production machines. The chainsaw example revolutionized small engine piston manufacturing, since several other engine manufacturers followed suit. The company was innovative in piston manufacturing techniques and had supplied hundreds of various piston process manufacturing machines and entire systems to customers around the world.

It was common for each of the companies to have a specialty or two that gave them an edge in that area, although they could still expect a serious challenge every time. That particular company specialized in machine tools for the manufacture of several other components as well, connecting rods and transmission valve bodies as examples. The company was sold by the founding family, then consolidated with another, then finally disappeared with decline of that company and eventual purchase of the remnants by a German company. That invaluable piston and other component machining expertise disappeared in the process. The other companies that disappeared had specialties as well, which were lost or at least diminished in the industry consolidation. You may recall from Chapter 3 that an estimated 80 million engineering hours were expended developing that expertise and were a resource available to exploit for future production advances.

That component-specific expertise was gained over years of competing with other machine tool companies, winning some and losing some, but incrementally advancing the state of the art with each exercise. All the gains made through the lessons of marginal early successes and painful failures and redesigns have lost their meaning. The subtle details of process technique and of the component response to material removal, tooling, and fixturing forces are very narrowly known or understood.

Even so, the loss of the network of companies competing with each other to advance the art is many times greater than just the loss of current levels of expertise. The trend to purchase component parts from suppliers, who in many cases have long-term supply relationships, further retards the rate of advance. Should they be concerned about the competition, which continues to exploit the capability and the ingenuity of multiple competing suppliers and/or special machine tool companies? You bet!

These are distinct losses for engine builders around the world, although it is doubtful that they all recognize it that way and will likely view this discussion as "sour grapes." The customer's current management doesn't seem to understand the special machine tool industry or appreciate it or its obstacles to reasonable success, which is ironic considering its historical impact on their own business. There is more involved and at stake than that auction target price. The current special machine tool purchasing "low price paradigm" blinds those companies to the real impact on their own organizations

In the 1960s and 1970s, auto companies were concerned about special machine tool capacity and would do things like place purchase orders to simply reserve capacity in machine tool schedules. They actually acquired significant equity in some special machine tool companies and made partnership commitments with others. Both were wrong, as the competitive aspect was affected, but it demonstrated their appreciation, at the time, of the importance of the industry to their own businesses. Those purchases and partnerships have since been reversed.

Standardization

Benchmarking to establish manufacturing process technique is viewed as a good concept, but by definition, in a competitive environment such as the automobile industry, it is already too late. Someone else is already doing it. By the time emulation could occur, the aggressive competitor may have implemented the next generation. The objective must be to implement processes that are generations ahead of the benchmark.

In a similar situation, American auto companies with numerous engine plants in the U.S. and in other countries are adopting a philosophy to standardize on "world class" process and hardware for all who produce the same part families. On the surface, this sounds like the right thing to do. It is not, for several important reasons:

- The lowest common denominator to achieve consensus on the approach to be selected is likely to prevail; probably a safe, conservative approach.
- Local manufacturing cultures and related pride are typically grounded in some logical local factors. Arbitrary change can create resentment, no matter which country or location is being asked to

change their ways. An example of this kind of cultural paradigm is the preference for rear-wheel drive cars in Europe versus front wheel drive in the U.S. Is either wrong? Local organizations must feel part of the process and be able to succeed or fail. The method of dictating based on a non-local concept removes local ownership and does not show appreciation or reward. Spirit and passion will be negatively affected.

- The dictatorial, centralized approach trains to not innovate locally and discourages local original thought and ownership.

- The "standard" approach freezes in time the entire corporate process, whether extremely effective or a failure, or anywhere in between. Any change, even fine-tuning, is a massive undertaking and will be discouraged.

- Different local processes encourage innovation with many points of entry for new ideas - friendly competition. Failure or less performance than expected is localized. A culture of Yankee ingenuity!

- It's possible to move to newly developed methods quickly in a limited way, one location at a time, each better than the last. Exposure to risk is limited, while staying ahead of fast on their feet competition who will be pushing the productivity boundaries. Solid communications between locations will leverage the different experiences.

- Each new local retooling will be the next generation rather than the entire battleship. In the "standardized" process, competitors can be several generations ahead before enough momentum can be built to scrap the old and build a new battleship.

- Worst, the world wide imaginative and creative resources of the special machine tool industry will be used infrequently and ineffectively. Remember also the effect of the single huge population on progress discussed in Chapter one.

Standardization and innovation clash. They are mutually exclusive by definition, and must be intelligently balanced and managed.

131

Personnel

There is an area in both the automobile industry and in the special machine tool industry that is troubling. In the latter, during the 1970s and 1980s, while the industry was consolidating (like others), the entry of fresh young technicians and engineers slowed to near nothing. In more recent times, the need for those people grew, as many experienced people had retired and the use of advanced technology intensified.

It is now difficult to attract the right kind of candidates for the openings and a significant experience/knowledge gap exists in many organizations. In effect, the staff formula mix has been seriously diluted.

As an example, to become a reasonable special machine tool designer will typically take six or eight years or more. A new employee can now become proficient with computer aided design (CAD) tools within a few weeks. To some, the distinction is not clear. The computer is a tool and has little to do with the designer's skill and talent. The earlier mechanical draftsman became proficient with a straight edge, triangles, and the compass within a few weeks. They are both only tools to be used. Either way, the number of years changes very little, even with the better tools. The exchange of the elder experienced designers for proficient CAD operators has confused some and created some difficulty. In addition, the retained older designers did not gravitate to CAD as readily as the younger ones. (Computer aided design tools, used expertly, are an invaluable resource and this discussion is not intended to minimize that in any way.)

When numerous companies and even entire industries, such as the auto industry, have decided to cut cost and "downsize," one of their first actions is to offer buyouts to senior employees. This creates a similar problem. The staff formula mix is seriously diluted, since those with the most experience go first. Their value to the company, in terms of performance, has not necessarily been a factor in the decision.

Health and longevity expectations have changed significantly in recent years. Today, a 60-year-old can look forward to many productive and fulfilling years, and many want or must continue to be active and/or earn income. What a shame that we do not let them and at the same time handicap the organization by the loss of their capabilities and contributions, a serious disservice to both.

"After he retired at age 60, Othmar Ammann designed, among other things, the Connecticut and New Jersey Turnpikes; the Pittsburgh Civic Arena, Dulles Airport, the Throgs Neck Bridge and The Verrazano Narrows Bridge. Paul Gauguin "retired" as a successful stockbroker and became a world famous artist. Heinrich Schliemann "retired from business to look for Homers legendary city of Troy. He found it. After Churchill made his mark as a world statesman, he picked up his pen and won the Nobel Prize for Literature at age 79."[25]

Ronald Reagan's second four-year term as President of the United States, arguably the most demanding job anywhere, began at age 78. What about Nelson Mandela? What do you think about the difficulty level of his job to govern and to take his country out of apartheid at age 71 after spending 27 debilitating years in a prison?

"Not only are more of us living longer (and often a lot longer) but we are not spending the extra decades twiddling our thumbs. Perhaps we're inspired by something maverick novelist George Eliot once put into words: 'It's never too late to be what you might have been.'"[26] or to continue to be fulfilled in a quality "life's work" for all "the reasons we work?"

For the most part, seniority should be its own reward and should not be the only factor in evaluating overall performance for position security or advancement. Those that feel that work is labor and have not realized the great benefits of the "reasons we work" could be among those chosen in a reduction in force.

The customers/clients of the special machine tool industry sometimes do themselves a very serious disservice by simply not challenging that industry in a fair and honest competitive bidding process to find better

---

[25] United Technologies Corporation, *A Message,* (Reprint from The Wall Street Journal, 1981).

[26] *Editorial,* (Remedy Magazine)

ways. There are many examples of dramatic breakthroughs that have resulted from a hungry competitor out-thinking worthy competition. The process, done well, can energize the imaginative, competitive passions of great companies from around the world.

Chapter 9:

Competition

Company Competition at the Level of the Individual

The free market system works because winning new business against worthy competitors is a constant and serious challenge, since all those who are competing want or need that business. Those organizations that succeed in winning and effectively executing that business prosper for themselves (owners and employees), and provide value for their customers, adding to the human well-being in general. The level of competitiveness depends on their motivation. A turned on organization, in which the "reasons we work" are facilitated and imagination, passion, and courage are fostered, will succeed on a reasonably level playing field.

In an earlier chapter, a point was made that in the effort to create a quality response to an inquiry, the special machine tool company in effect pitted their people against their competitors, people in a virtual head-to-head contest for the winning solution. The individuals in the client organization, auto companies for example, responsible for acquiring equipment are in the same situation with respect to their individual counterparts in the competing auto companies. The goal of these people is to enable their company to develop and offer high value products more effectively than their counterparts in worldwide competitor companies. Obviously, procurement of the most effective tools (processes and hardware) will be instrumental in achieving those goals. It is not always seen in that light, as a bureaucratic environment and a low initial purchase price auction can obscure the real objective at the engineer and buyer level. They may not appreciate the threat posed by their aggressive individual counterparts.

The outcome of the procurement exercise includes how well the equipment is selected as well as how well it is managed. It is also what is learned in anticipation of the next product generation and its associated production equipment. It can be a great idea, but if it is not imaginatively operated or managed as intended and the lessons learned from it are not absorbed, it would not advance the process.

Anticipating the toughest, most tenacious competitor creates the effect of bringing out the best competitive response from all of us. The effect is apparent in professional athletics. Consider a typical preparation of the second place team for a game against the first place team versus the preparation for the game against the last place team. Less serious competitors can elicit less preparation. In a business competitive exercise, all the toughest competitors are in every game with just one winner, so the maximum effort is necessary every time if we hope to succeed.

Some companies are able to create an atmosphere that develops energy based on beating the competition, even specific competition, as in professional sports. It is a cheer leading exercise that injects energy and momentum. We must anticipate the toughest every time out. The perception of being the weakest could draw a less competitive offering. A laid back, low-key, and non-flamboyant profile makes some sense. The anticipation of a tough competitive exercise every time will keep us sharp and on our toes and growing from the experiences.

No matter how creative and how well a solution excels in its ability to solve the problem at hand or to satisfy a customer's inquiry, there will always be a better one. The buyers/users of the special machine tool products must understand that and also realize that their competitors from around the world could beat them to that better solution. If the objective is low price, without full consideration of the other important tangible points of value, the benefits of competition cannot be realized.

Market Distortions

There are four market distortions that have adversely affected competition and the free enterprise system, the global special machine tool industry, and very likely other industries as well. They may be seen by some as "sour grapes."

First, it is fact that the people involved in the acquisition of special machine tools in industrialized nations outside the U.S., even Canada, can be extremely nationalistic and will buy locally, if at all possible. It's true even if the equipment is to be used in a neutral country, and even if it might be supplied from a base in the buyer's home country belonging to the American competitor. In many cases, they will admit that preference.

Ironically, their counterparts in the U.S. demonstrate a preference to buy from abroad. The lure of foreign travel, with amenities provided by the sellers, is very attractive. The perception that a particular culture, German for example, can provide superior quality or craftsmanship is a potent sales pitch, whether or not it is true.

In recent history, currency exchange rates have been a very serious handicap for purely American companies. Those companies with multinational bases can mix content for favorable pricing tactics, which is the case with several of the European companies competing in the U.S. market. Some have manufacturing facilities in South America and supply bases in Eastern Europe, further enhancing their cost structure. This is a fair approach and American companies should consider that strategy as well.

Second, many industries, including the American special machine tool industry, practice customer gratuity relationships. That is, they do a certain amount of entertaining, such as lunches, dinners, golf and sporting events, all considered appropriate and proper. One special machine tool company went well beyond the norm in that practice which significantly distorted the competitive landscape.

In many cases, the product to be produced or the equipment being proposed is in its early stages of design and subject to change along the way. The prospective corresponding change to the production equipment in process can present an opportunity, at some later time, to recover a financial shortfall that may have been precipitated by a low bid to get the order.

That company cultivated close long-term relationships within certain customer organizations that had ineffective checks and balances. It was given an understanding of what the initial price had to be to get the purchase order. The anticipated change orders would then include the difference to make it profitable, as well as to cover a compensation of some sort to the customer or individual. Surprisingly, this practice continued for many years and had a significant negative effect on the industry.

There is no question that orders would have been spread more evenly across the competing companies if the procedure had been done properly. The same company appeared to have a policy that let other companies invent new processes and hardware and they would follow. It makes sense, as it would not be intelligent to take risks when not required. So not only did the undeserving company get the business, but manufacturing progress was retarded because it was not necessary to push the envelope to get the business. The purchasing companies were at least as responsible for the practice as the supplier and had the most to lose, both in terms of the total cost and in the technology level of the acquisition. Competing companies may have been able to acquire more advanced equipment for less cost, the products of a truly free market.

Third, major automotive companies have adopted the principles of auction based target pricing - the concept of one individual. He was originally revered as nearly god-like in cost saving results at the highest levels. On the other hand, he was run out of the automotive business for his dishonest and subversive activities.

The auction process focuses on driving the quoted price down to the lowest possible level without full regard to the value of the solution being offered and the life cycle cost impact on the product to be produced. To have the best chance to receive a purchase order the machine tool companies must offer the low initial cost solution, regardless of other numerous considerations that could affect the cost of the product component to be produced and its quality.

This concept polluted the free enterprise bid process for companies that adopted that practice in acquiring special machine tools. The high integrity sealed bid (solution and price) concept that not long ago was respected and fostered technology advance for the buyer is now passé.

Fourth, the European, and specifically the German, business climate regarding knowledge of and respect for the machine tool industry provides the basis for favorable treatment from their local government and financial community. Companies that have survived in Europe during very difficult business years and now are very difficult and aggressive global competitors may not have survived in the U.S. Terms for capital needs can be more favorable and flexible to the local companies, and the governments have provided various kinds of support and protection.

These market distortions have a definite affect on the pitch and yaw of the playing field in the special machine tool industry. They distort the free enterprise system. Worse, they dramatically degrade the result for the buyer in terms of the true value of his purchase and the advance of manufacturing technology for the entire manufacturing community.

Once put into effect, newly developed ideas and concepts from real competition eventually become public domain, so everybody everywhere loses when less competitively-inspired solutions are utilized. Further, the really great ideas temporarily protected by patents force competitors to find an even better way or they lose big time right now and for years to come - the free enterprise system in action!

Evaluating the Offering

The inquiry examples for production equipment in Chapter 2 draw responses that can vary for content and concept all over the map. The buying organization's evaluation and recap of the various offerings must be summarized to provide focus on the bottom line--in other words, the cost effect on the product to be produced over the life cycle of the product. It is a given that before getting that far, the approaches being considered are viable and capable of producing pieces to the required specifications at the required rates.

Initial purchase price is a major factor and the most visible one, but by no means the only one. Typically, budgets for capital equipment purchases are established for specific new product programs. The path of least resistance for evaluators is to use initial price against budget numbers as selection criteria without a realistic value put on operation and other factors.

There are several other factors that have a cost impact on the parts to be produced: the ongoing cost of perishable cutting tools and their holders; the frequency of and time to change tools; the cost of other perishable components; the number and type of operators required; the hardware and software robustness (reliability/maintainability); projected preventative maintenance and repair costs and downtime (lost production cost); floor space consumed; material handling; and loading and unloading requirements. The cost of these factors plus the equipment purchase price and its installation costs translated into cost per part produced over its forecasted production life is the real bottom line for comparison. All of these factors can vary widely in the various solutions offered.

It's possible that one supplier can offer a viable approach with the least hardware and thereby, it is the least expensive alternative initially. However, the other factors in their operation could make the parts produced more expensive than that of a more expensive initial purchase. It is difficult to comprehend how the auctions that are currently being used can properly consider all those factors.

The Suppliers' View

A special machine tool company reacting to an inquiry may decide on its response as follows:

- Does the inquiry fit our profile?
- Is it a customer that we can relate to?
- Is it a production part that we can comprehend?
- Has a reasonable time line been provided?
- Are the production part tolerances and the acceptance criteria realistic and achievable?
- Does a resulting order fit a specific space in our backlog satisfying the prospective customer?
- Who are the competitors and what are their strengths and their backlogs - are they hungry?
- Are the competitors recognized for their experience in production of this particular part in comparison to us?
- Are the competing companies likely to use a traditional approach?
- Are there other factors involved, such as; is this a new customer that we want to impress for the future; or a new product to be produced that will provide an advantage for the next time?
- Is this customer really interested in the best approach or is he just looking for the rock bottom lowest possible initial price, wasting our imagination capital?

The people in any organization that are the imaginative stars are a finite resource and are carefully assigned to important prospective opportunities. To use that invaluable resource on a customer project that will not fully consider the unique content and its real value or in a situation where that content will likely be disclosed to competitors prior to any commitment makes no business sense. Do you suppose that those buyers have thought this through, or do they care? The low price auction and the culture of Yankee ingenuity are incompatible!

In other words, can we and do we want to win this order? We can accept the inquiry and agree to provide a quotation, or decline to quote the inquiry.

Inquiry accepted and the quotation process follows:
- Establish concept to be offered. This is when the imagination contest occurs – to secure a purchase order against clever and aggressive competitors.
- Establish selling price
  - Standard formula?
  - Inflate for negotiation (auction)?
  - Under formula for competitive position?
  - Over formula based on the perceived value of imaginative concept chosen to be offered?

It is surprising to many that normal formula pricing for the American special machine tool industry puts profit in the 5 percent of the selling price or less - after all costs, but before taxes- range.

On a level playing field, special machine tool companies that are part of larger and diverse corporations are less likely to participate on the same competitive plane as privately-held companies, even though passionate and knowledgeable people may manage them. This is primarily because they are subject to the close scrutiny and the business practices of the corporation and perceived stock price influencing factors. They may not have the freedom to compete head-to-head against freewheeling, independent companies on landmark projects.

A board of directors representing typical investors, institutional and otherwise, will not always be sensitive to or have sentimental feelings about a business that isn't consistently living up to minimum financial performance standards. As in some examples cited earlier, it is possible to ride out several years or several rounds of competitive exercises, winning some and losing some, but sooner or later a pattern develops and impatience at the corporate level causes changes.

The International Marketplace

Four prominent German special machine tool companies now compete from U.S. facilities. Three are in the Detroit area and one is in central Ohio. They are all excellent companies technologically and are worthy competitors. In a sense, they are American companies, as they have American employees and local facilities, although much of their product will come from abroad. Two have facilities in low labor cost countries in South America, from which components, sometimes major components, can be supplied. Those companies receive their direction from Germany, take tactical advantage of currency exchange and low costs from their various locations for their American customers, and serve their transplanted auto customer countrymen with a nationalism advantage.

Two of the companies purchased American companies and two began operations from scratch after supplying initial orders from abroad. The largest is a part of a very large German conglomerate and has been well-known in Europe for many years. Initially, it acquired a small U.S. company and then several years later, it acquired the larger group. It purchased the U.S. special machine tool company that was the result of the consolidation of seven other companies. It is now the combination of eight original special machine tool companies not including any that may have been incorporated in Germany. Also included was the U.S. standard machine tool parent of the acquired company. A mix of people from different locations manage the various parts of the company, but all are machine tool people. While they will have some work to do to re-establish themselves, laying claim to the best aspects of all the new combinations of their pieces, they will likely survive and prosper.

Prior to the American special machine tool industry's downturn, the leading companies did well in the U.S. market, did a substantial amount of business with vehicle producers in the former Soviet Union, and had some success with the auto companies in Japan. Ford and General Motors had significant manufacturing presence in Australia and utilized primarily American special machine tools until about the mid 1980's. More recently, they have dealt with Japanese machine tool builders, likely because of favorable currency exchange rates and their geography.

Several American companies established important bases of operations in England and Germany, partly to overcome nationalistic biases. Two of the larger companies of the time were never successful and withdrew, but others did very well until the decline of the industry starting in the late 1990's.

Today, the European companies have the advantage of serving the U.S. market from Europe, the U.S., or their low cost Third World bases, depending on activity levels and exchange rates. In some ways, they have exchanged competitive positions with the American companies of the 1970s, except that the foreign locations are very receptive to their products and presence.

Until very recently, the Japanese special machine tool companies have not made a serious run at the North American market, except to supply the Japanese automotive transplants extensively from Japan, which by itself a very significant level of business. Some American companies did participate successfully in the first transplant programs as well. More recently, the American auto companies have encouraged the Japanese machine tool companies to become involved and as a result they have supplied major machining systems. It's too early to assess the results from either side of the new relationships.

Why are the current German industry leaders apparently succeeding, while the American companies are struggling and failing? The basic reasons are outlined in the foregoing discussions. The most important of these was the leadership, ownership, and passion vacuum remaining following the passing of the entrepreneurial generations. It's entirely possible that the other obstacles to success could have been overcome with the kind of vision and passion demonstrated by the entrepreneurs before the stresses imposed by succession concerns (the find a way mentality).

In Germany and other industrialized countries in Europe and in Japan, the machine tool industries and manufacturing have generally been much higher-profile and are still held in high esteem by the general population, business, and government. With the higher profile, the "rust belt" image was not present. The businesses continue to be highly respected and are able to attract capital and fresh employees. The industry's apprentice programs have always been highly respected and their ability to attract highly educated engineers and others is good as well. The knowledgeable and committed people do gravitate to the top jobs.

There have been a substantial amount of business failures in the European machine tool businesses as a result of severe economic recessions. Failed business arrangements with Eastern bloc countries in the 1980s and 1990s also negatively impacted that industry. Some companies have survived with the help of government and banking institutions. Those recessions played a major role in those companies' decisions to have a serious presence in the American market, the largest market for those products in the world.

It is apparent that the German auto industry's efforts are parallel to their machine tool industry's efforts. The three major companies have a significant new presence in the U.S. and in the Americas in general, and of course, one has become the owner of the former Chrysler Corporation – now DaimlerChrysler. Where do you think these German auto companies will go for their machine tools for their U.S. operations? It may take some time, but it is likely that even the formerly American division will develop that same bias.

In Chapter 3, it was stated that the American auto industry was far and away the leader in mass production during WWII and that the special machine tool was integral to mass production. The American auto industry is still a leader in mass production, but with difficult competition coming from all directions. The European auto companies are pressing hard, and aggressive Asian competition will continue to be a serious threat ahead. Great, high value products, very innovative manufacturing, and passionate and responsible people create a very serious threat.

Special machine tools are still integral to high volume production and to any specialty manufacturing assignment. At present, the American companies are far less prominent than they once were. The American "find a way," Yankee ingenuity mentality, along with the client industry's recognition of dynamic value in their purchases in a free market, can regenerate it. It may turn out to be with the next generation of companies!

Chapter 10:

Government and Pig Farmers

A recent newspaper article talked about a Midwestern state's pig farmer's problem. The problem was that the cost to raise, feed, and bring pigs to market left little or nothing for the farmer. This offered that state's U.S. Congressional representatives an opportunity to serve their constituency by finding a way to get a subsidy that would sustain the farmers through the crisis.

This particular action may or may not have been the right thing to do, as all the background information was not presented. The point is that there are many such subsidies in place and the Congressional people who promote them get credit and votes. It is likely that there are varied opinions on their legitimacy and their urgency, but you can bet that we will all pay for these subsidies one way or the other in higher taxes, reduced government services, or increased prices. We will also pay for it in the loss of the benefits of competition for overall value - price and quality. There is not much need for Yankee ingenuity to find better ways to improve the product or its cost through genetic engineering or other imaginative approaches when they are subsidized.

It is well-known that family farms and more recently, the larger corporate farms as well, historically have received a great deal of sympathy and ongoing subsidy aid from state and federal governments. The natural question that arises is, what establishes the priority for farmers in modern America in comparison, for example, to the family-owned or otherwise small special machine tool companies or any other business, especially the small entrepreneurial businesses? Are they of greater strategic importance to Americans? Is it possible that farm ownership could involve foreign investors or owners and could involve U.S. government subsidies as well?

The subsidy actions parallel welfare and organized labor examples where ongoing aid or protection can be a strong incentive to not fix a problem. They also blur John Quincy Adams's view of individual liberty, and they encourage dependency. In most such cases, the market is sending a very strong message and needs to be heard. The message is typically that someone else is doing a better job of providing value and deserves to get the business.

A story told to children about the eagle with a broken wing says that it needs to be put back into the wild as soon after recuperating as possible, or it will soon lose its ability to hunt and to sustain itself and will have to be supported for the rest of its life. Subsidies can have that same kind of effect.

It also seems that in the modern global free market environment, subsidies can interfere and be a source of friction between trading partners. No matter how you look at it, the free market is being manipulated by the use of subsidies except to fix a short term tilted playing field problem.

Loan Guarantee Subsidies

The Chrysler Corporation received help in the form of a government loan guarantee that likely saved it from the scrap heap. The company was saved and to its credit, it paid off its debts well ahead of schedule and became a model of efficiency. It was prolific in its well-received new model introductions and set records for profitability. It was so successful that it ultimately became a part of the premier German Auto Company through acquisition.

At about the same time, another American manufacturing company of significant size also received a U.S. financial bailout, also through loan guarantees. One of its problems was that its manufacturing capability was out of date and inefficient. It used its bailout leverage to buy new standard machine tools to fix that problem. Foreign built machine tools were selected. We cannot second-guess that choice, as it could easily have been the right thing to do, even courageous under the circumstances, as it hopefully was a value based decision.

That company, like Chrysler at the time, succeeded mightily, and today, it is sitting on top of the world. It has been saved from the scrap heap and is successful probably beyond its management's wildest dreams.

The first point is that if you had to guess, what kind of machines do you suppose would have been purchased had this been a German company or a Japanese company that needed help? Is this a level playing field problem?

The second point is that in retrospect, it seems that these were the right things to do as the employees, the owners, and the customers all benefited. There were however, negative points involved as well. The bailed-out companies were failing, likely by their own shortcomings, over a period of time. How do you suppose the competing companies viewed these actions? They may have been the finest, most upright and competitive companies that you can imagine, foreign and domestic. Did they deserve to be handed a major handicap for doing everything right? What about their employees, owners, and local communities? All the successes of the bailed out companies came at their expense.

These are difficult questions to answer, but the free market was artificially manipulated, as in the case of other subsidies. The point is that intervention is a very complicated and sensitive thing, and it can have an impact far beyond what may be a noble cause or in some cases, a political agenda. Some jobs may be saved or created, but many more may have been negatively affected. Worst of all, the system that causes advances (progress) for all suffers a setback with each such tampering. In addition, do all the competing companies now feel that if they get into serious trouble that they will be bailed out by the government also?

Tax Abatements

Another kind of subsidy that can affect competition is also a form of competition, ironically enough. It is competition between locations.

States or other political sub-divisions offer tax incentives to companies they would like to attract to their location. It could be a foreign company making its entry into the U.S. market, or a U.S. company moving to a more favorable labor market with an added benefit of a tax advantage.

The local governments realize that they are in a competitive market and must be very aggressive regarding tax rates and other business climate issues to attract new business. They want to create jobs and all the peripheral things that result, which is a good thing. The balancing act involves locally attractive labor rates, so that it is attractive to business but will also improve the local prosperity level, all things considered.

The real problem is that the basis for a very competitive tax incentive is that other local businesses and citizenry must subsidize it, at least initially, and could resent unequal treatment. Their share of tax must somehow offset the incentive. The tax incentive for a foreign entrant or a migrating company seems to advance the consumer's interest, as these companies should become more competitive. It will help to make any company more competitive, but at the expense of others.

There are Southern states that have attracted a large number of businesses, including well-known foreign auto companies, with tax abatements. Is it possible that those states may now have a tax system marginalized by those subsidies? Their ability to stay current in infrastructure, highways, and public services in general in a population expansion environment may be strained. The new jobs promised are low paid on average and may not bring the anticipated prosperity to those states. It could be a difficult situation when the auto market cools or if a deep recession occurs. There was substantial worldwide excess auto capacity even before construction of these Southern facilities, so competition will be very tough for the foreseeable future.

A greater reason for success of these transplanted operations will be their own ability to develop a totally green workforce, which is free of Detroit paradigms, into one that is world class effective. The labor costs will be better for three reasons: a competitive hourly rate, the workforce attitude (a version of a W3 environment), and minimal retirement benefit burden in the early history, a major handicap for the Detroit companies.

A basic truth involving any subsidy is that by itself, it will do nothing to improve the overall basic operation of any enterprise. In fact, the subsidy can cause the recipient management to defer or to forget about more serious efforts to refresh a substandard competitive base. The new lower local labor rates and tax burden will help but that may mask a more serious deep-seated problem.

Subsidies can have serious negative effects on the provider, distort the free enterprise competitive environment, and impose penalties on other local businesses and citizenry. It can create low paying new jobs in one area at the cost of presumably better paid jobs somewhere else. It is likely that to make the move in the first place, except to enter a new market as in the case of the auto companies, there were problems difficult to solve any other way, surely a symptom of more serious problems in competing. Even attractive subsidies are no substitute for Yankee ingenuity.

The Special Machine Tool Industry and Government

The American special machine tool industry has always been a low profile business, perhaps by the design of the original entrepreneurs. The industry, probably because of its history of private ownership, has had a macho kind of pride that would resist asking for help from anyone for any reason. This fact has been a handicap to the industry, just as that low profile hurts its stature in attracting financing and prospective employees. It also obscures it from the government, except of course the IRS, which then has little reason to be aware of unfair competition or of events or circumstances which in other industries and other countries could trigger assistance.

Our form of government forces the representatives to be sensitive to their constituency's needs and demands. The American special machine tool industry is relatively small and virtually invisible to the general public (the voting constituency) As a result, the representatives feel little pressure to pay attention to it, if they are even aware of it. The time required to really make a difference would net few votes in the next election. From their view, it is a poor investment of their resources.

It is very clear that in a Midwestern state with a concentration of agriculture or of automobile industry, the representative knows what the constituency needs or wants. It may be to organize a subsidy designed to protect an industry through a short-term crisis. It may be to stop dumping on the part of a foreign competitor, or it may be to get through a year of drought or flood. It should be in place only long enough to diagnose and fix the core problem. Then the industry affected should return to the competitive free enterprise environment at the earliest possible time, like the eagle and for the same reason.

In the late 1970s and early 1980s, there was an effort to develop machine tool industry awareness within the government by an industry group from Rockford, IL. The effort was based on the need that would exist for manufacturing expertise in the event of a national emergency. Mobilization would require a healthy American special machine tool industry. There were government hearings and meetings held over a considerable period of time. It was not taken seriously and eventually, the effort simply faded away as the government seemed to lose what little interest it had. Some of the companies involved in the effort were affected by the industry decline and different priorities required their people's efforts elsewhere.

This was an example of the fact that, unlike Franklin Delano Roosevelt, recent governments have not recognized the industry for what it is or what it does. In fact, it is unlikely that more than a handful of today's members of Congress even know of its existence, let alone the part it has historically played in U.S. manufacturing.

Ironically, the special machine tool company most actively participating in the effort, a division of a major conglomerate, was euthanised by its parent not long after this effort collapsed. That company was actually the combination of several old-line American special machine tool companies consolidated as a division of the conglomerate and working to regain respectability. The corporation lost interest since the returns were less than they could realize in other investments. It included a very respectable German division, which was first sold to its employees, then went public, and is now one of the successful German companies competing worldwide with a significant North American presence.

Other countries' machine tool industries are much higher-profile and are recognized by and have the sympathy of their populace and governments. Their governments paid attention to the plight of those businesses, not unlike the U.S. government with agricultural and other subsidies

Germany will take extreme measures to help or to save companies who find themselves in trouble. The Germans are really trying to save jobs. It makes little difference that the company may have been totally mismanaged or even taken advantage of by previous management, or that it may really bring little value to the market. In conjunction with banking institutions, they give financial aid, they forgive loans, they even take an equity position and then find a buyer for the business for whatever the market will allow and subsidize the difference.

It's a kind of reverse law of the jungle: the survival of the weakest or unnatural selection. It is a disservice to the successful companies, the future of the industry and, above all, to the consumer. The consumer will ultimately pay the cost of the subsidies and at the same time will be expected to buy the products of businesses that are less than competitive on their own merits and probably offer products of lower value.

The American special machine tool industry has not and is not now asking for government help or intervention. What should be expected, however, is that the local representatives of government, the elected Congressional people, should have an in-depth knowledge and understanding of the businesses in their constituencies. It should be expected that in the event of an apparent problem that involves foreign competition that may somehow have an unfair advantage that the representatives be available to council and advise. The advice should involve finding the way to fix the core problem and not designing a subsidy, thus perpetuating and compounding the problem. This would be difficult if the representatives did not have a working knowledge of the industry and its impact on other industries. In other words, there should be a natural and ongoing brother - sister kind of dialog without specific obligation either way.

Monopoly

In recent years, a great deal of media attention has been focused on anti-trust concerns relating to a perceived monopoly and its alleged anti-competitive practices. Microsoft Corporation is the defendant in the legal action, a very large and well-known high tech company. This company is responsible for innumerable advances in its field of expertise that has been a great benefit to its many customers, worldwide, and, by their success, to the American economy. The facts that gravitate to the general public make it very difficult to judge the merits of the concerns. Their competitors have been instrumental in the Justice Department's decisions to pursue the legal action. Obviously, these competitor companies are less successful than they may want to be as a result of the offending company's successes, right or wrong. They would like to neutralize a growing competitive disadvantage.

Monopoly: Exclusive control by one group of the means of producing a commodity or selling a service.

In an entirely different area, there is a real and long-standing obstacle to fair competition that is not pursued as a legal issue. In fact, it is defended by more than one government agency. While it is obvious, that issue is obscured by the paradigms of another time. Sixty years ago, stormy labor management relationships: required tight controls to maintain order, insure some level of fairness, and to minimize disruption; the job of the National Labor Relations Board (NLRB).

That long-standing obstacle is present in various segments of business today, but it is exemplified prominently in the American auto industry. An organized labor strategy is used whereby labor agreements with the three major auto companies are pattern-bargained. That is to say, a target company is selected based on which has the most to lose at the time of the negotiations and a very demanding set of negotiations ensues. (The three contracts expire at about the same time.) This arrangement is accepted by management and defended by the NLRB. The threat of a crippling work stoppage strike has a very coercive effect, as the competing companies will not be threatened until the first comes to terms.

Once the agreement has been completed, the next two, in a sequence favorable to labor, are negotiated and agreements are reached based on the pattern established in the first. If a work stoppage was necessary, it would have likely only affected the first company, the one with the most to lose.

So now that the pattern agreements are in place, what are the effects? The single largest factor influencing value to the buyers of the products produced, the cost and effectiveness of labor, has been taken out of the local competition equation. By virtue of the cloned agreements, which include direct labor costs per hour, fringe benefit costs, and work rules, this major factor has been neutralized competitively. Doesn't this breach, at least in spirit, the Justice Department's mandate to preserve free market principles? Isn't this a monopoly on the part of organized labor?

What about all three companies' competitive posture against foreign competitors in the modern global market as well as in the domestic market? Is this comparable to the "single huge population" versus the "small isolated population" problem suggested in the second chapter?

The opportunity for any of the big three companies to use creative methods in their organizations to provide their people the fulfillment of the promises of the reasons we work resulting in a creative and "turned-on, ultra competitive organization" is lost. The major components producing value, the imagination, passion, and courage of the human element, inspiring Yankee ingenuity are missing or impotent in that setting.

In the American auto industry prior to the late 1960s and the serious entry of foreign auto companies into the U.S. market, the neutralization of competitive labor costs had little meaning that would be obvious to the general public. Added labor costs and perpetuated or even increased inefficiency costs associated with new agreements for all three companies would simply be passed on to the consumer. They would not affect competitive positions in the eyes of the consumer.

Today, however, the competition, both global and transplanted, plays by their own set of rules regarding the effectiveness of their workers and the way they are treated.

The rapidly evolving, revolutionary technology, and global business environments have changed the competitive landscape. Are the needs of the modern global business environment obscured by the paradigms of times gone by? Considering both examples, are today's anti-trust laws current and comprehensive enough? Are they able to scrupulously assess the fairness and benefit aspects for all involved, including that of the general public? Are the individuals in organized labor getting the good deal they think they are?

What about the global free market? Major league hardball must be expected. It will require the most competitive posture possible, including effective, imaginative, and fulfilling use of labor, of all collar colors. The selection of the most effective tools possible and their imaginative and effective use by labor against clever and low labor cost companies and countries are fundamental to survival, success, and any hope of security or prosperity.

In thinking about this problem, it is noteworthy to realize that the many thousands of the Big Three autoworkers are in a closed-shop environment. That is, all must be dues-paying members of the union and follow that culture without recourse. The states involved do not have "right to work" laws. In an open shop, those in "right to work" law states, the labor chemistry could be significantly different. A major part of the work force could opt for a different life's work atmosphere. It may be one that recognizes the realities of drudgery versus fulfillment and success in an increasingly difficult marketplace, which is the only real job security.

Further, those competing businesses with no workforce affiliation to organized labor would seem to have even greater opportunities for flexibility. They could be quick to respond to creative and aggressive competitive maneuvers or, better yet, to lead, requiring response from competing companies. We don't have to guess which auto producers in the U.S. today fit which description.

The National Medal of Technology

"The National Medal of Technology is the highest honor bestowed by the President of the United States to America's leading innovators. Enacted by Congress in 1980, the medal was first awarded in 1985. The medal is given annually to individuals, teams, or companies for accomplishments in the innovation, development, commercialization and management of technology, as evidenced by the establishment of new or significantly improved products, processes, or services.

The primary purpose of the National Medal of Technology is to recognize those who have made lasting contributions towards enhancing America's competitiveness and standard of living. The medal highlights the national importance of fostering technological innovation based upon solid science, resulting in commercially successful products and services."[27] In principle, it is truly a Yankee ingenuity-inspired effort.

From 1985 through 1998, 96 medals had been awarded. In the first year, 1985, one of the medals was awarded to John T. Parsons and Frank L. Stulen for their development and successful demonstration of a numerically controlled machine tool. The other ninety-five awards involved almost as many industries. The award for the numerically controlled machine tool was a major milestone for that industry and, in fact, it precipitated a revolution in the special machine tool segment and in numerous others.

The President of the United States gives the award after a recommendation by the Secretary of Commerce, following evaluations made by a nominating committee. The nomination itself can come from anyone, but must follow strict nominating guidelines.

It is significant that the one machine tool award occurred very early in the history of the medal, as the special machine tool segment had already begun its decline. The decline in the early stages was not as noticeable in business volume as it was in a general fading of the industry stature as

---

[27] National Medal Technology home page http://www.ta.doc.gov/medal/ - US Department of Commerce

evidenced by difficulty in attracting new people and by all the consolidations that were taking place. It was being painted with the "rust belt" paintbrush.

It is entirely possible, even probable, that some of the many developments and innovative processes and products that have come out of the American special machine tool industry during the medal years would have qualified for consideration and that medals would have resulted. The industry leaders had their heads down fighting for survival and there was no one else to wave the flag on their behalf. It is very likely that local, state, and federal government representatives did a certain amount of cheerleading on behalf of some of the 96 recipients

This was not a problem that affected business particularly, although that kind of recognition could have been a shot in the arm when things were really tough otherwise. The medal dry spell has simply underscored the decline and further obscured the industry from the view of the public and the government.

Chapter 11:

Lessons Learned

Situation Summarized

Yankee ingenuity, powered by W3 energy, is a rich American legacy and has been a major influence in the great history of the United States, in its dominant role in the global economy and in the well being of its citizens. It continues to be an important factor in the success of many American endeavors, enterprises and industries. Actually, ingenuity from around the world and the free market forces or modern natural selection, as suggested in Chapter 2, facilitate accelerating advances all around us. While that may seem to be a threat to some, it is really the force that will drive all worthy competitors to do better and better. It is the cause of the rapidly improving human well being, that is, it is life quality and longevity; technology and medicine.

Manufacturing industries, which started it all in the U.S. in the era of the American Revolution, are being selectively de-emphasized in the U.S. today. That appears counter to the direction of other economic powers in the world and in Third World countries as we enter the third millennium.

Stock market analysts will say that the U.S. is shifting from a manufacturing economy to an information economy. The obvious question is, is that a good thing? We can't eat, wear, or drive information. Obviously, if manufactured goods are no longer produced as they once were by Americans and they are still in increased demand, others are producing them.

The computer, information and telecommunications industries have captured the imagination and attention of much of the American general public. Business, financial, academic, and government organizations are distracted by them as well. Those still-evolving industries do play a vital, even dominant, role in the current and future success of the national economy and global economy.

The stock markets have even differentiated between the two areas of the American economy. One exchange is said to contain the "new economy" or "tech stocks," while another, or at least its major index of "industrials" contains what is referred to as the "old economy" stocks. While the tech stock index has seen a dramatic reversal from its "irrational exuberance" in its early history, that business segment is expected by some to eventually set the pace for business once again.

As the transformation in the U. S. favoring the new economy continues, the focus and intensity in some of the traditional industries decline. Even their relevance is sometimes questioned. The tougher and tougher global competition in those industries continues to gain momentum, adding to that distress. The transformation, at times, seems to be headed toward exclusion of important industries from the American economy.

The new industries provide a continually expanding array of exciting products, which by their nature are designed to enhance and support other products, services, endeavors and enterprises. The new products are not the tangible things that we surround ourselves with and with which we create our comfortable and abundant lives. Our homes, our transportation, our furnishings, our appliances, our highways and of course much more are those tangible things. Industries such as agriculture, chemistry, and medicine are life-sustaining and critically important as well.

All of these fields of endeavor in the "old economy" are being greatly enhanced by the still evolving "new economy" industries, but cannot be replaced by them, nor should they be diminished by them.

The "old economy" industries are the major users (customers) of the products and services of the "new economy" industries. The new economy industries in turn are prominent among the users (customers) of those enhanced products. The two economies are not mutually exclusive, but are mutually supportive and mutually dependent. It may not be represented exactly that way by financial analysts or financial advisors since their agenda relates to share valuation, which can be dominated by glamour factors and many times, is very transient.

There is another underlying conundrum at work here that distorts the free market environment. The prospect of getting rich overnight in escalating new economy overpriced stock (price based on perceived growth potential) has reduced the appetite of the investment community for the challenge and risk of the old economy stocks including manufacturing of all kinds. (At the time of this writing the trend is now stalled, at least temporarily, by a significant stock market correction. Many believe the trend will resume.)

This circumstance has diluted the energy of the Yankee ingenuity culture in the old economy industries; the ones which hopefully will continue to provide our life - quality and life-prolonging advances. The magnitude of the gold rush to the discoveries in the new economy has diverted the creative talent, energy and capital resources away from the apparently less rewarding old economy. That emphasis will likely continue through some maturity of the new economy and the resulting rationalization of earnings expectation, both real and share value growth.

This area is far to complex and extensive for discussion in this particular book, but it will certainly be a major factor in the success of the composite future American economy.

While other nations continue emphasis on the traditional areas along with the newer ones, has American business become so preoccupied that it has taken its eye off the ball? There is little doubt that while certain other economies, the Japanese, for example, have faltered, they have continued to quietly do their homework and ironically, they may have become the next sleeping giant.

The German manufacturing aggressiveness is obvious in the automotive and the machine tool industries and in others as well. It is depressing to read about the American automotive companies' deteriorating market share. The lists of models enjoying sales leadership certainly must be discouraging to the American companies as well. In the early days, it was the small, inexpensive, fuel-efficient foreign cars that attracted value-conscious buyers and they ultimately became the sales leaders.

More recently, foreign manufacturers are dominating the luxury and intermediate car market segments as well. Their most recent targets are SUVs and light trucks, where their progress is obvious. This must be depressing to the American shareholders, management and labor alike. Financial security and job security for many may be in jeopardy.

The strength of the overall auto market in a sense masks the real problem because the total sales volumes are still quite high. The market share problem would become more apparent if the market strength cooled and the total sales numbers returned to earlier levels or if an extended recession occurred. It could be a traumatic awakening! A softer market cannot support the current industry capacity.

It's also depressing to understand the fate of the American special machine tool companies that have been discussed. The special American heritage, Yankee ingenuity, while as potent as ever in many industries, is not being focused as it could and should be. The automobile and machine tool industries are obvious examples. In reality, what has become known as the "rust belt" is only rusty because we think of it that way. It has not had the attention that it deserves and that the other industrialized nations give it.

The American special machine tool industry leadership of the 1970s has declined to a small fraction of its one-time market share in the global industry. The particular difficulties that the industry has had to deal with may be a problem for other industries as well. (A brief early discussion pertaining to the American standard machine tool industry segment did not elaborate on its status. Unfortunately it is in trouble as well due to many of the reasons discussed).

"But while the industry (auto) as a whole frantically is trying to cut costs in an effort to offset slowing sales and rising incentives, the underlying conclusion of the study (KPMG LLP) is that most excess costs have been wrung out of the supply chain, and future gains will depend on technological innovation."[28]

Interpretation of technological innovation: The technology involved in the enabling of competitive production of innovative new products - in other words advanced processes and production hardware.

Where will those advances come from? In their internal cost reduction efforts, the American auto companies have come to rely heavily on their special machine tool suppliers for specific manufacturing engineering capability. In recent decades, they purchase entire manufacturing systems competitively from a single supplier. The best original combination of process and hardware are selected.

Formerly, they would devise and manage the entire detailed manufacturing process as a whole themselves and purchase individual or groups of machines from multiple suppliers. Without that need, that capability is likely to have been greatly diminished. You will recall points made earlier concerning countless advances made possible by the collective imagination and energy of the special machine tool industry realized through the free market competitive bid process.

During the period of decline of the American auto industry and the special machine tool industry, the competing industries from Japan and Germany, as you might expect, have gained strength along with market share. Their machine tool suppliers have benefited from their robust market strength. They benefit in being selected by American auto companies for machine tools as well, certainly with help from very favorable currency exchange. In the meantime, the once great source of imagination and energy of Yankee ingenuity in American special machine tools is underutilized and languishing.

---

[28] Drew Winter, *Study: Detroit to Continue to Lose Share for 5 Years.* (WardsAuto.com, January 10, 2002).

The environments within the American special machine tool industry and in its marketplace have become unfriendly to the culture of Yankee ingenuity. The opportunities for technology advancing value for the user organizations are seriously diminished by retrogressive procurement tactics. This environment discourages free thought with oppressive structure, rules and other paradigms. The industry transition from entrepreneur to public ownership, along with management succession stresses, has weakened it considerably. These are serious problems in any free enterprise system for both supplying and procuring companies.

There is a concern that high tech products are advancing so fast that they overtake their consumer industries' ability to apply or incorporate them effectively. The products in question are not just the finished PC, modem, scanner, wireless communication devices, or software, but the base components and concepts that will be integrated into the products of others with custom software.

It is reminiscent of the relationship of cutting tools and the machine tools that depend on them: boring tools, milling cutters, etc. In alternating phases of technological advance, machine tools must wait for advances in cutting tools to be able to take advantage of their own capability. Then the opposite is true: Cutting tools wait for technological breakthroughs in machine tools for their product capability to be effectively utilized. There are exciting new technologies originating in the special machine tool industry itself.

The possibilities for application of accelerating technology are limitless. In the past, customers and competitors were the forces driving advances and innovation in most industries. The demand for innovation now comes from a third direction. That force is the pressure to apply the rapidly advancing support technologies. It is intensified by the other two, as well as by others, including Wall Street.

Industries that have been thought of as mature are no longer mature, as they find themselves facing new horizons. In some cases, their paradigm registers have been or will be set back to zero. Remember the Swiss watch industry following successful demonstration of the Quartz watch (it threatened the entire Swiss economy). Many gears, springs, and jeweled bearings, the defining components of a watch are no longer required. Can it really be a watch?

The point is that these industries, and especially the American auto and special machine tool industries, must possess abundant, passionate, and imaginative, engineers, designers, technicians and their enlightened managers and owners to even survive and then to prosper, now more than ever. They must out-think, out-apply, and then outdo their global competition to succeed. It's major league hardball!

The term "agility" (discussed in Addendum One) has particular meaning in this discussion. To standardize and freeze in time the use of a certain technology in product concept or in its manufacture is likely to be imprudent. It may seem at the time to be the latest and greatest and most flexible, but technological advance tells us that a quartz watch may soon be available. While most Swiss companies were looking for better ways to apply and make gears and bearings, someone "found a way" to eliminate them entirely.

The technology advance worldwide translates into intensified competition. There is a clean sheet of paper, or a clear Computer Aided Design monitor screen, in front of a lot of people these days, in other words, opportunities and challenges. What a great and exciting time in our history for Yankee ingenuity. We must find a better way to use it effectively.

Lesson 1: The Global Free Market

The global free market is the force responsible for the accelerating progress: life quality and longevity in the human well-being. Progress has reached levels that would have been inconceivable even in recent history.

- The basic culture of the free market must be understood, respected and its challenges accepted by buyers and sellers alike. It defines the progress curve for us all. The courage to follow one's imagination, to accept challenge

165

and to risk failure in that marketplace as Eli Whitney did two hundred years earlier is fundamental to the realization of the fruits of human imagination: progress.

- Enterprises of all kinds, but the special machine tool industry in particular, including the many small companies serving manufacturing companies around the world, are the small, isolated populations. They are the incubators of innovation. Their constant search to find better ways and provide imaginative and competitive dynamic value to consistently win new business against worthy competitors brings rewarding opportunities to their employees, their owners and their customers and means prosperity and "progress" for all.

Lesson 2: The Human Element

The great benefits of ones life's work are only implicit in American society and in other truly free societies. They are not guaranteed. That is, they are available to all for the taking, but they are obscured to many and may become more obscure to others over time. It may be that it is not always thought through or taught in our early years. Focus may be lost because of perceived difficulty of circumstances, or comfortable routines becoming rigid paradigms.

Some environments teach (perhaps by example) that mindless labor and drudgery are normal, thus deflating ambition and positive perception of the future. Other environments preach government and organized labor control philosophies, which can profess to shelter us from risk and insecurity or offer unrealistic security without corresponding contribution. They will discount the need to be adventurous in accepting challenge and risk. Still others will not see the need because of birth status and inherited material wealth. In that context, you may remember that the majority and the most rewarding of the rewards (W3) are non-monetary, far too great a penalty to pay even for the advantaged.

The demands of the special machine tool industry, its competitive nature, and the need to outsmart and outperform worldwide competitors on every single piece of new business fosters the culture of Yankee ingenuity. The only security offered is the success of the efforts in a very challenging marketplace and against "take no prisoner" competitors. Facing challenge and risk energizes courage, imagination and passion and awakens latent capability.

The human element is everything. It is the source of free thought and imagination that create and it is the energy and passion that implement. It is responsible for all things that make a difference-progress.

The competition for attracting, retaining, and encouraging passionate, imaginative and free-thinking team members must be as intense and as aggressive as the competition for winning new business. These efforts should include:

- Develop and maintain an environment that makes fulfillment of the promises of "the reasons we work" a top priority for all.
- Foster and develop the courage characteristic, as a practice.
- Eliminate any psychological barrier between white and blue collars, empowering all equally under their leaders.
- Provide and require ownership, a stake for responsibility.
- Establish a succession plan that assures that industry reared people are in line, with appropriate ownership for all management positions

"The 1998 edition of the EWC Engineering and Technology Degrees survey covers data from 340 schools with engineering programs and 284 schools with engineering technology programs in The United States."

"According to the EWC, between 1986 and 1998 , the number of students receiving bachelor's of science degrees in engineering declined by 19.8 percent to 63,262 nationwide while the number of students receiving bachelor of science degrees overall increased by nearly twenty percent over the same period of time...In Connecticut where the 'Connecticut Yankee' has long been a symbol of ingenuity and inventiveness, only 533 students received B.S. degrees in engineering last year, one third as many as the state graduated in 1986," said Torpey, Chair of the 20 year old AAES. "How can a state that considers itself 'engineered for high performance' believe that it will be able to fuel technological innovation without an adequate supply of engineers to provide the spark of Yankee ingenuity."

"As our society becomes increasingly dependent on its engineers to maintain our nation's economic, environmental, and national security, our community has a responsibility to improve the nation's 'engineering literacy' as well as a responsibility to encourage and inspire our nation's youth to consider engineering as an exciting and rewarding career," said Torpey. "As Motorola CEO Gary Tooker said two years ago, 'The nations that lead the world in the decades to come will be those that encourage creative people to become engineers.'"[29]

The concerns expressed by Mr. Torpey and Mr. Tooker are even truer today and must be taken seriously. They present a major handicap to American industry and in particular to the special machine tool industry. The second lesson of Yankee ingenuity underscores these concerns and must add fuel to the efforts to encourage young people to become engineers. However, the discussions on recruiting along with those on collar color in Chapter 6, if taken seriously, can help to mitigate that problem.

Lesson 3: The Business Environment

A virtual "no rules environment" is necessary for the culture of Yankee ingenuity to flourish. No rules in the sense that free thought and imagination are encouraged, facilitated, and unrestricted and are not subjected to repressive standards, traditions, or paradigms.
This applies to both the technical side of business and to the organizational aspects. The rules governing fair play to all, business integrity, and morality, remain in their proper place at the highest level. But remember: standardization and innovation are mutually exclusive.

---

[29] *Press Release,* (The American Association of Engineering Societies, January 11, 1999).

Lesson 4: Managers and Owners

The "employee owners and owner managers" concept is the one that solves several problems associated with ownership, management, succession, etc. in the American special machine tool industry. It also satisfies the insightful observations of the venture capital executive who put employee passion and executives' specific expertise at the top of his "must have" list. The problems regarding outside executives as managers following an entrepreneur passing without effective succession plans are eliminated as well. Most of all, the culture of Yankee ingenuity has the best chance to flourish: preservation of the small, isolated populations effect!

The Buyer/User's Role

By now, the value of the American special machine tool industry to us, the final consumer, is apparent. Its role in continually improving consumer product value and innovation is a key to progress. It would seem reasonable, then, that the buyers/users of the special machine tool products who manufacture the consumer products must also recognize that value. Their shrewdness in capitalizing on the competitive, creative resources of the industry can pay large dividends. It must be done in a fair and professional way with as few rules as possible. This by itself would give them a key role in restoring the industry to its former level of efficiency and respect, from which they will be the greatest beneficiaries.

The decline of the industry, the benefits lost to date, and the prospect of further loss in these capabilities is difficult to quantify. Further loss is possible, as the two large remaining American special machine tool companies seem to be vulnerable. They each have characteristics that also played a part in the difficulties experienced by those companies that have not survived. In addition, there are numerous small but effective closely-held companies who will ultimately face the same problems that hurt the companies discussed. These are the larger companies of tomorrow which will be the basis for a resurgent American industry.

This should not be interpreted as a plea for mercy or for treatment other than fair, professional, arm's length relationships with favors to none or all, whether, foreign, or domestically owned.

The return to a true sealed competitive bid process is essential. To be truly effective, it must be done with the highest integrity and mutual trust possible. The objectives must be for the buying organization to get the best dynamic value time after time and for the suppliers to have a reasonable opportunity for success. It is free but fair competition that drives technological advance. The bidder's price and concept offered must be held in confidence. The offerings should be compared based on true value over the life cycle of the products to be produced, typically requiring a buyer and a manufacturing engineer to work together closely.

Progressive Payments

An old problem for the industry and for many of its customers from the opposite side of the same problem is "progressive payments." It's apparent after discussion in earlier chapters that the capital-intense nature of the industry presents it with several problems. The shame is that the working capital part of the equation has little to do with any company's real capability to innovate and supply very large and sophisticated quality machine tools and manufacturing systems.

The main problem is that the company's financial resources will establish a line above which business cannot be accepted. It will not be able to finance multiple large projects over an extended period of time. This is regardless of all of its other resources and capabilities. In other words, the best-qualified company, having won an order in all other respects against large and difficult competition, with an exemplary imaginative offering, cannot be given the order. It results from the fact that its financial resources, relative to working capital, limit it.

Selecting that supplier can be a distinct advantage to the buyer, except for the hassle that is required within his company to arrange progressive payments. This situation is a major barrier to further progress for the offering company, as it will precipitate a less deserving competitor's gain. The offering of the next best competitor will be a less desirable one. That means, in effect, that both the buying and selling companies are negatively affected.

There is something wrong with this picture. We all know that it is possible to provide progressive payments as many companies and industries do it all the time. The real problem is that the policies and systems in place, when designed, did not recognize this prospective need and as a result, the process is extremely complicated and a major burden to those involved. You will remember that formerly single process steps typically were purchased as opposed to the entire systems that are common today.

These policies and systems now need to be redesigned since they hurt the buying company as much as the seller. The buyer's choices become more limited in a leveraged opportunity to save cost and advance his technology level. The cost for the use of the money in a progress payment arrangement is typically returned to the buyer, as a discount based on current interest rates.

Without progress payments, the entry barrier to prospective spin-offs or other new prospective businesses, the small, isolated populations, is much higher. The number of significant new companies entering the business, during the years of history discussed, can be counted on one hand. There are other factors, especially the human element, but finance is the big one.

Standardization

The multi-location, multi-national manufacturers made a mistake trying to standardize processes and methods worldwide, losing opportunities to innovate locally, energizing and motivating their local team members while respecting their cultural differences. It's also the opportunity to incorporate emerging technological advance locally without betting the whole farm and to make updating to subsequent technology generations a huge task.

"This is the way that we in Detroit have decided you should do it" mentality wipes out the free thought, the "no rules environment," and the local feeling of ownership.

The reason to utilize advancing technology is to provide capability to manufacture components either previously impossible or impractical, or at a newly-specified higher rate or quality level or lower cost. A current example: the machining of wing spars and similar parts from solid billets of aluminum versus fabrication with fasteners was not practical until developments by one of the American special machine tool companies made it so.

The objective for the buying company is to be better able to compete in the global marketplace against ingenious and very aggressive competition and to stay on or exceed the progress curve. To freeze a manufacturing process to the extent that change becomes difficult, time-consuming, and expensive, even if it's ahead of the progress curve at the time, will surely be a handicap against aggressive, fast on their feet competitors. It's especially true with the technology curve as near vertical as it is.

Technology on the Production Floor

Higher technology levels imply easier operation and maintenance to some. Hence, no higher skill levels are required for its use or maintenance. This is simply not the case!

The 1950s automobile provided the basic functional transportation capability and would function as well today as then. Many of us could do fairly large and essential diagnostics and repairs on those cars in our own driveways or at the local corner gas station and consumers were required to do so on a fairly regular basis.

Of course, there are many enhancements for comfort, efficiency, safety, and emissions on the current model cars. The frequency of these kinds of repairs on today's sophisticated cars is far less. However, it takes educated and trained technicians with sophisticated computer diagnostic tools to accomplish what could be done in our driveways in the old days.

There are some modern devices that are available to us today in our cars and our homes that require some thought and patience to understand and operate. The use of PCs and related devices are proliferating, and even getting full use out of a VCR requires following fairly complex guidelines. Doing things wrong can damage equipment and erase valuable, essential data or programs. Life in general, as well as in the workplace, has become more complex, but most of us can learn to do things differently and correctly with desire, attention and concentration.

This all has to do with the operation and maintenance of the more and more advanced equipment once it is installed in the buyer's facilities. By virtue of the technological advances now incorporated and the even more advanced ones on the horizon, the operator and maintenance skill, education and training level requirements, and organized labor work rules are a serious problem.

For the advanced systems to live up to their true capability, leveling the playing field against low labor cost countries, the people operating and maintaining must be highly trained, skilled, multi- disciplined and, above all, they need to be motivated and to be "owners" (take responsibility).

If the production floor mentality normal to many American auto plants persists, the new capability will not be realized. In fact, the effect will be to become less effective than in the past due to the increased complexity to operate and maintain devices by untrained and unmotivated people. Specifiers of equipment understand the problem and may buy less sophistication, since in the short term they will be more effective. In the long term, they will surely lose.

The alternative is a current trend in the automotive industry. It is to establish long-term relationships with large parts suppliers, rather than to fight the battle of changing their local culture. Some of those suppliers push to acquire the most advanced systems practical and have the desire and capability to operate them at their maximum efficiency levels. They will make significant gains and will be the winners, while the others, the losers, include employees.

These part-supplying companies are consolidating, creating much larger companies with much greater capability. They expect to supply complete assemblies of parts from vehicle seating and bodies to entire braking systems and potentially transmissions and even engines. While this approach creates other potentially less effective operations from an agility and innovation point of view, it seriously challenges the typical automotive manufacturing environment in productivity and product cost.

Lesson 5: Procurement and Use

This is probably the most important lesson of all. It's an old lesson of the free market system and must be continuously re-taught, as it is often too easily forgotten.

Let the free market system work!
- Value based procurement choices
- Value based (W3) "life's work choices"

Exploit the system to the maximum advantage of the buyer in finding the best value. As unlikely as it may sound to some, this will also be to the maximum advantage of the seller in a truly free market. The seller must go all out on every competitive exercise and win orders based on its ability to out think its competition. It will grow to be a better company because of it, benefiting all. Encourage alternative solutions and evaluate them shrewdly and fairly. If Jeremiah Wilkinson could revolutionize the production of nails in 1776 and a special machine tool company could revolutionize the way small pistons are manufactured in the 1970s, think of what might be done in the early third millennium with technology exploding all around us, all in an effective, competitive environment.

That creative energy will thrive only in an environment that rewards on the basis of fair and intelligent use of the free market system. Fair exploitation sounds like an oxymoron, but in fact it is the key to the system success. Fair in this context means protection of the bidders' proprietary submission – selling price and concept offered and value based choice. Yankee ingenuity will flourish in this environment.

- There is an imperative need for a universally fair, high-integrity, competitive sealed bid process.

- Special machine tool purchase decisions in the global market of today must be value based without regard to country of origin or other non- essential issues. These decisions should consider the uniqueness of their local manufacturing environment, their motivation level, knowledge and pertinent cultural factors.
- Redesigned internal financial and legal mechanisms of buying organizations to make progressive payments available in appropriate cases will greatly benefit all concerned and will help perpetuate the industry through the reduced financial entry barrier. The cost to the buyer = zero!
- Work rules and the motivational levels in many American manufacturing plants must be modernized for advanced technology to have any meaning. Very competitive parts suppliers, domestic and foreign, will do that for them, over time, if they do not. It may be their only opportunity to survive and prosper against low labor cost worldwide producers.
- Lean production and agility, discussed in addenda, must be rationalized for, or overlaid on, the mission of any organization to determine their most effective formulation. They are not a "one size fits all" commodity.

Two puzzling questions regarding current special machine tool procurement philosophies;

- Why do the procurement practices/tactics of Jose Ignacio Lopez de Arriortua, who is alleged to have betrayed the extraordinary levels of trust vested in him by more than one auto company, still prevail? He failed to effectively exploit the American competitive free market and the magic of Yankee ingenuity for his company because he obviously didn't understand them. He prioritized personal power and instantaneous low price gratification over the true dynamic value of an

175

acquisition as it's reflected in the cost and quality of its own organization's final products. He infected the American free enterprise system with a low price paradigm virus. (Low initial price by itself has little meaning in the assessment of the value of an intellectual product.) The demonstration of his own lack of personal integrity certainly should call into question the practices he institutionalized and are still in use by some companies today, a decade later.

- Why standardize a manufacturing process and its hardware for multiple and international facilities, disregarding local culture, local contribution, and responsibility, and effectively freezing the entire capability in a time when both product and production technologies are, or should be, advancing exponentially?

The Government's Role

The U.S. government, unlike those in Germany and Japan, has very little current knowledge of the special machine tool industry, and therefore, has even less sensitivity to its plight and its impact on other manufacturing. The local Congressional representatives may be aware of the existence of some of the companies, but apparently have less than a true appreciation of their role in American manufacturing. The industry has always been very low profile with regard to government involvement probably by the design of the original entrepreneurs. Their preference was to keep the government out of their business, exercising their independence.

There has not really been a serious cry for help and that is not the case now. It seems, however, that other industries enjoy a high profile with their representatives and get help in the form of subsidies when market conditions head south. It at least shows recognition and sensitivity.

In the global markets that exist today, it would be impossible for the relatively small special machine tool companies to make a case with a foreign government that a competing company might be enjoying some unfair advantage.

Lesson 6: Governmental Relations

The special machine tool industry needs friends in government, perhaps the Congressional representative, who can counsel on matters that will help to maintain a level global playing field. They first need to become educated on the industry and its value to the country. It is also likely that a better job can be done to take advantage of the best technology in the world that may be available and applicable from the American space and defense industries (our tax dollars) through those representatives.

One Final Lesson

An overseas auto company purchased several special machine tools from one of the smaller American companies. All had unique features, but one was a first and a very complex and sophisticated machine. In addition, the order was accepted on a very aggressive schedule to allow the customer to hit a window of opportunity in his planning process. The aggressive schedule was accepted as a favor to the customer although he would not remember it that way when the chips were down.

The machinery had to arrive just as the new concrete floor was ready, following the removal of the old machinery, revamping of other portions of the process equipment and then be ready for production shortly after. The components to be produced had not changed, but the new process would be more cost and quality effective. In other words, the old equipment could still produce the components if something went wrong and if their removal was done in a way that preserved that capability.

In the final stages of completion of the new machines, it became apparent through the combination of the aggressive schedule, their complexity, and the technology stretch involved, that the committed schedule was becoming a very serious challenge. Still facing some unknowns because of the newness and not wanting to jeopardize the customer's production deadline, it was suggested to the manager that care be taken in the removal of the existing equipment and even some

177

contingency planning be done to restart it on a temporary basis in the event of a problem.

The position taken by the customer's management was surprising. Their feelings were that any effort to provide a safety net for potential problems provided a reason to fail, and in fact, the likelihood of failure increased because of it. The safety net philosophy could make failure to achieve objectives an acceptable outcome which can then become ingrained in the culture of the organizations involved. Another effect can be that the bar is lowered in setting objectives for the next time out. The customer proceeded to scrap the existing machines.

Many of us, in that kind of situation, would have welcomed the comfort of that safety net. It seems so logical to have your backside covered in the event of a problem when the stakes are so high. However, that comfort will dilute the energy and urgency to "find a way," the Yankee ingenuity solutions, for the unexpected challenges and problems encountered along the way.

While this is a business or industrial example the principle fits almost any setting and their challenges in everyday life. Should we always take the safe approach?

The machines were delivered and put into production, late by a tolerable margin, with no small amount of difficulty for all, but finished vehicles were produced as planned. Another effect of the success is that the bar may be raised in setting objectives the next time, by lessons learned and by confidence gained and become more competitive in the process.

The wisdom of that philosophy was not as apparent in the heat of battle and with its stress and extra effort as it was following the battle.

End of Story

We started out talking about just a couple of kinds of grapes but have covered just about all the fruits and vegetables in the market. It has been a recollection of experiences and thoughts coming out of nearly fifty years involvement in the special machine tool industry. You are to be congratulated that you have come this far, or is this a case similar to the tree falling in the forest with no one to hear it. Did it make a noise? Is anyone still reading? Does it make any sense? Does the significance of Yankee ingenuity and special machine tools in our history come through?

The intent in doing this book was to illustrate the reasons for the decline of an industry not normally visible to the general public in America. It is, however, an industry that has served an important function in the American success story. Even more important are human factors that have been the foundation of that industry and without doubt, the historic American success story. An important question that we must ask ourselves is; are those foundation factors eroding more broadly in American business and industry than we realize?

Much of the subject matter has not been special machine tool specific and probably applies in many industries, vocations, and professions, and as stated in the beginning, there is some opinion involved; would you believe it?

Several recurring terms have been a large part of the discussion: imagination, passion, courage, rules, the "find a way" mentality, the W3 "why we work" logic and the exercise of our individual liberty. They are the "foundation factors".

The culture of Yankee ingenuity is in stark contrast to some existing environments in American industry today. Merit and recognition are bad, the "suck up" mentality, "the blue collar glass wall," "the auction process," "this is how we in Detroit expect you to do it," and repressive rules and paradigms are the antithesis of the culture of Yankee ingenuity.

Passion for what one does, as in life's work, is the most powerful tool anyone could have for whatever his or her goal happens to be. It can overcome or sidestep any obstacle and when teamed with imagination, courage and the reasons we work, W3, will work wonders – these are the "sweet grapes."

In which of the three ingredients of human personality does passion originate, and can it be influenced by life's experiences and can it be acquired through training and mentoring? While there is likely a genetic influence, passion, like imagination may be awakened or instilled through training, mentoring, and especially by example. In America, the fulfillment of the promises of the reasons we work is ours if we really want it. Remember that the personal liberty that we as Americans, Yankees, all share is personal power. This is something John Quincy Adams told us, but it takes our courage to make it work.

The American economy today is the largest and most effective in the world, just as it was during WWII. The manufacturing segment is a smaller percentage of the total than in those times. While the national well-being is in excellent condition, the challenges from the global free market in manufacturing causes some nervousness. Those capabilities are essential to our well-being, our needs, comforts and our recreation and in most other ways that you may think of.

The challenges the special machine tool industry faces today in its own recovery, along with the need to apply accelerating technology advances, are difficult ones. The culture of Yankee ingenuity must be reawakened in the industry and in its client organizations, the users of its products. The need to be imaginative and innovative in using technology plays to the industry's strengths and along with heeding lessons of the past will be key to its re-birth.

The American special machine tool industry can sustain itself, grow, prosper, and be "sweet grapes" intense. It will need the help of those who benefit the most from its capabilities and services, its customers, by letting the free market really work for it.

Those readers who have interest in some of the technical aspects of special machine tools, especially in their application in the auto industry, stay tuned. The addenda cover certain aspects of the technology and additional historical perspective.

Addenda

Addendum 1:

Lean Production and Agility

Lean: Lacking in fat, not using any more resources than necessary

Agility: Nimbleness, responsiveness to demand

"Lean production" and "agility," two words re-coined in the early 1990s, were perceived by many as almost magical answers to puzzling questions about American and European eroding automotive manufacturing prowess against that of the Japanese. The case for that feeling was justified in some areas. In other areas, different pictures emerge that create questions regarding that conclusion.

Lean production integrates several concepts that intend improved organizational and methods philosophies. The dynamic work team, product design for ease of manufacture (lean design), just in time inventory (kanban), continuous improvement (kaizen = unrelenting and imaginative search for better ways and waste elimination), and agility are key among those concepts. Lean design, continuous improvement, and the dynamic work team concepts are like motherhood and apple pie, and will elicit little argument from anyone except possibly organized labor.

Lean design incorporates the concept of "simultaneous engineering." S.E. is the process whereby product engineering and manufacturing engineering are brought together at the earliest time practical to insure that the product detail is developed around the most effective way to produce it. This concept is particularly advantageous when an organization is free of paradigms that could encumber the imaginative search for the best solution, whatever it may be. Machine tool purchasing policies and practices can fall into that category. It's clear that the low price paradigm, "auction process," discussed earlier has distorted the process for the search for the best manufacturing solution in some procurement organizations.

The dynamic work team, an important advantage of lean production, is defined loosely as one made up of all levels of staff, multi-disciplined, with those actually adding value to the product given the maximum number of tasks and responsibility. They are provided access to a high level of management information and real time data to manage output and achievement of their goals. The key implied meaning contains some of the attributes of the reasons we work, passion, resourcefulness, and a kind of "find a way" Yankee ingenuity. Organized labor has serious difficulty with this concept, for reasons discussed earlier, and could be an obstacle in the achievement of real leanness.

Other aspects of lean production such as "just in time inventory" and "agility," which obviously have an important place in production, do not fit all circumstances into which they are force fit.

Just in time (JIT) inventory typically means that long term commitments and special relationships with suppliers exist that are counter to the small isolated population concept discussed in chapter two. The idea of minimized inventory certainly is a good thing especially if it could be achieved while maintaining the highest level of competition for value in the supply of that inventory at the same time. Some would question whether JIT actually works or is the safe inventory level burden simply shifted to the supplier?

Agility relates to all aspects of a manufacturing enterprise. Simply stated it is the ability to be nimble and effective in influencing market trends and demands by staying a step ahead of competition, or responding effectively to competitive challenge. Offering continuously competitive customer perceived exciting variety and dynamic value is essential. The products incorporate emerging technology both in the offerings themselves and in the technology to produce them.

Customer perceived variety and its competitive position in automotive products are increasingly being strategically planned to be achieved above the platform level. The number of platforms within many auto company's portfolios are being reduced significantly based on designs that permit ready adaptations of variations of the upper body and trim, to name some.

Examples: a very popular retrograde small utility vehicle on an existing traditional small car platform and new SUVs on existing truck platforms or on auto platforms. Recently, a decision was announced that would reduce the number of platforms between two cooperating and partially jointly owned auto companies from 29 to 13. The same kinds of efforts are occurring throughout the global auto industry. Obviously there are some changes required to existing platforms for new vehicles and subsequent generation platforms will come along as well.

These changes are being made to improve costs and competitive position expecting that the total final production volume is shifted, not reduced. So rather than production volumes going down on platform components, because of variations going up, production is actually going up on the fewer number of base platforms. Platform/chassis, components include suspension, axles, steering, brake systems, engines, transmissions, and differentials (drive train). The vast majority of the machined components in an automobile reside in the base platform.

Higher platform volumes on fewer variations of those platforms will achieve automotive product variation. Agility will have the importance ascribed to it for the various final product models identified above the platform level. It will not at the platform level and below because the high volumes should make dedication with reasonable flexibility a more practical and leaner approach. Terms such as mass production, economies of scale and dedication, thought by some to be archaic, are in fact as important as ever in the high volume arena (platforms as the example). Flexibility to accommodate typical changes or improvements, even later generations in the high volume dedicated systems is normal;

however, a machine meant for crankshaft production will never make cylinder heads.

The main goal in consolidation of platforms is increased speed of new product to market, nimbleness in shaping the market with exciting new products and in response to market demand. The very risky but successful example of the retrograde styled vehicle to get a competitive advantage may not have been tried if it had required a new platform as well. The byproducts of cost savings, particularly engineering, and risk reduction in new products are significant.

Nonetheless, we must look at agility or flexibility in the manufacturing of the thousands of different components in the platform.

Agility, when applied as a manufacturing philosophy, has been defined to have a more far-reaching meaning. When it is applied to a more specific component part production machining system the classic definition applies. In general it can be interpreted to mean that it would be nice to have the ability to quickly convert a set of hardware from the production of one component part to a different one. The intent being to satisfy changing demand between component parts and/or to redefine that part easily to respond to competitive pressure or to convert to the next generation product quickly, inexpensively and dependably. That hardware choice should not in any way restrict the product designer in his efforts on succeeding component generations. Of course that capability should not add to the cost or complexity of normal operation beyond its perceived flexibility value.

Agility without other complications is very desirable for obvious reasons. The problem is that it would have to be a magical solution if it could be practically applied to all machined components regardless of their production rates, their forecasted life expectancy, or their similarity, and not negatively affect their cost and quality.

At the two extremes of the concept, the term takes on somewhat different meanings. In a job shop type of setting where short run, on demand, production is the rule, serious flexibility is not only nice to have but a requirement to be successful. The capital investment in flexible machine tools, machining centers, can be expected to pay dividends.

In a high production environment the reasoning is not nearly as clear since there are numerous factors involved that impact the economics and other effects of the decision to incorporate agility.

For example, the projected stability of the design of the component to be produced and the forecast production volume requirements in relation to the useful life of the hardware are key factors. In other words will the equipment be worn out or approach technological obsolescence before its flexibility can be effectively utilized? A machine tool system producing 150 components per hour (normal in an American automotive environment) two shifts per day will make 500,000, (allowing for some inefficiency) identical, complex machine cycles in one year. An engine and transmission combination will typically have a production life of from five to ten years. During that time, of course, significant improvement changes will be incorporated.

Addendum 2:

Matching Centers Versus Special Machine Tools

Some believe that standard machining centers can be used effectively in a high volume production assignment to exploit their versatility. There are no doubt cases where this approach is practical, and in fact it has been done effectively for a considerable period of time. There are several factors that will bear on the advisability of that approach.

Producers of standard machine tools and machining centers are product designers, typically not manufacturing or process engineers. These are distinctly different areas of expertise. They do their best to incorporate the features that appear to be the most desirable in their marketplace. They would like their product to be all things to all people. They then mass-produce those machine tools for distributor sales and for inventory.

There are a few machine tool companies that have both standard and special machine tool divisions. Obviously, they would like to be able to take maximum advantage of all possible synergy between the two. Presumably each group is being staffed optimally for their individual business plans for their products and would not necessarily be able to sustain a prolonged period where one side or the other would dominate without some sacrifice.

The machining center engineers typically have not specifically applied their products to production assignments on a regular basis. Their normal marketplace, predominately jobbing environment shops, have their own application groups since that is the nature of their business and will reapply the machine tools numerous times in their useful lives.

The standard machine tool marketplace includes producers of every kind of component imaginable and normally at volumes much lower than in the typical automobile context. They typically have built in numerous "nice to have" features which would never be used in a high production environment. Note: It is a contradiction to the definition of lean when machining centers are used in high volume assignments, as there are resources (features) in the machines that are not used effectively or not at all.

So an important question arises: In a high volume setting, who does that application engineering work, or more importantly, who takes ownership of all the manufacturing process sequence steps (could be hundreds for a single component) for specific assignments requiring numerous machine tools.

As discussed earlier, there are many cases where a process step in the production of a high volume component can only be done marginally on standard machine tools or maybe not at all which obviously is seriously limiting to the component product designer. This is opposed to special machine tools that can incorporate highly specialized operations giving the product designer complete freedom to do things that may have never been done before.

One of the major points in this entire effort has been to emphasize the importance of Yankee ingenuity, the "find a way" mentality and the no rules environment to facilitate them. When the use of standard machine tools is emphasized or mandated to take advantage of mass purchases or for a questionable approach to achieve flexibility, the universe of manufacturing advancing opportunities becomes severely limited. Nearly as much creative thought is required to adapt the standard machine tools but with far less opportunity to optimize the process, exploit new machining ideas and provide the product designer greater flexibility for product innovation as well.

For example, the specifics of establishing the step by step manufacturing process for a complex engine component is extremely critical and is only learned by exposure and experience. The manufacturing processes for cylinder heads, connecting rods, cylinder blocks, pistons, and crankshafts, typical engine component parts, and their response to fixturing and tooling forces and the relief or imposition of material stresses are entirely different from each other.

When material is removed in a specific process step, creating or removing material stresses, and applying very significant tooling and fixturing forces the component itself will react by, bending, twisting, shrinking or expanding. In addition the cutting tools wear and the feature being produced will vary as well. These changes are either predictable by experience or unpredictable necessitating allowing for adjustment capability of some kind. It may even be by "in process" real time monitoring and feed back to a custom and maybe automatic adjustment capability.

The operation sequence is critical, as those operations that complete a particular component feature must follow those that may create unpredictable results. Obviously the finishing operations must remove all variables from previous operation sequences. There are hundreds of process steps in the production of many engine components, many of which will cause non-text book results.

In many instances, the tolerances on feature characteristics are very challenging and can be in the two or three ten thousandth of an inch range (.0002") or even more challenging. Add to that the statistically formulated acceptance criteria and the challenge is seriously multiplied. It's apparent that the application of an informed process sequence plays the major role in component quality and cost.

An important complication of this concern is that some in the American auto industry have, over time, reduced their manufacturing engineering capability through attrition. This was partly because of the conscious decision to buy entire process capability from one special machine tool supplier. The cylinder block example of a fourteen transfer machine line with hundreds of individual operation steps demonstrates the reduced effort required if the selected supplier can manage it. The former method involved multiple suppliers awarded individual operations on a piecemeal basis. In that approach the process and hardware conception, coordination and control responsibility remained with the buyer. The coordination of those multiple suppliers, particularly considering the need for a well thought out process sequence over hundreds of steps, is a very serious challenge requiring significant engineering staff. That kind of capability, historically, has not been available in the standard machine tool company's organization.

Transfer machine: Multi station in line special machines that perform two different operations in each station. Each cycle advances a full complement of work pieces one station producing a completed piece (planned operations) at the end of the machine, every cycle. These machines may either be totally dedicated or flexible.

A basic difference in component processing between the machining center flexible approach and a transfer machine approach (with or without flexibility) is that in the transfer concept the component remains in each process step for a very short period (24 seconds at 150 parts per hour) and then is transferred to the next step.

The significance is that it is located, clamped, and supported as required for one or two specific process steps to be performed. The fixture (component holding device) has only to provide clearances around the component for those particular cutting tools. It is designed to resist only the specific forces involved providing the best chance to minimize

188

distortion. In addition any distortion or stress relief changes that may occur are immediately allowed to relax as the component is released following that operation and relocated and reclamped for the following operation minimizing any impact. The machining sequence will be planned to put the distortion producing operations in a place that can be accommodated by the process as a whole.

In the machining center approach, the component is fixtured on a machining center while as many operations as possible are performed before moving to the next machine. There are as many duplicate machines as required to make the prescribed production volume. The component locating, clamping and supporting device must be able to support the component against numerous and various types of cutting tools and provide clearance for them all around the component. This clearly compromises the basic fixture structure and the component process integrity. In addition component stresses and distortions that may occur during any or all the operations in that single fixturing remain until all are complete and the component is released. As a result those distortions are cumulative, adding to the unpredictability, and are reflected in later process steps which must be considered in the overall process formulation.

You may recall that in earlier discussion it was stated that the real product of the special machine tool company is intellectual in nature (the entire comprehensive component manufacturing process) and the hardware is simply the manifestation. The question of who takes responsibility for the entire process is a very important one. The 80 million engineering hours of experience accumulated in the 1960s, 70s, and 80s by the special machine tool industry, discussed in Chapter 3, was an invaluable experience and knowledge base, and is now greatly diminished and fading.

The product designer would like his component to be as lightweight and as small as possible to fulfill the needs of its product function in the most effective way. Typically, he will seriously resist making allowances for size or strength that could be a requirement in its production. Therefore, the manufacturing process, that is the work locating, holding and supporting devices and the cutting tools, cannot impart forces that will distort or otherwise make the outcome other than perfectly predictable on a component designed only for its own functionality.

A non-magical answer that seems to satisfy many of the concerns expressed above on when or how to apply agility in high volume production such as in an automotive context follows.

If we think of a production facility producing certain components for internal combustion engines, it is not hard to visualize a line of machine tools dedicated to production of each of its component parts. Presumably the basic engine and its current and future prospective variations will be produced for some time to come, typically five to 10 years, and will be produced at least at a rate of 400,000 to 500,000 per year. We should be able to presume also that the next generation engines in the years to come will still incorporate those same basic components although ostensibly significantly different.

In the past, specific special machine tools have been dedicated to a particular group of operations within a broad total sequence of operations and could have extensive flexibility. Presumably many of those same or similar operations will continue to be required on following generations of the component involved. It is practical to design and build custom machine tools that will do those specific operations, even those not practical or possible on standard machine tools.

It can be done in a way that provides flexibility to do those operations on the new, but corresponding features on part variations and on future generations. This is a machine designed by special machine tool engineers, manufacturing engineering and process engineering specialists, who accept responsibility not just for machine tool functionality, but also for the effectiveness of the entire production process. It will incorporate whatever capability is required for its assignment, but will not be all things to all people. The flexibility can be extensive or narrowly focused based on the particular component features to be produced and the degree of flexibility desired in anticipation of future component variations.

A good example of this concept is the crankshaft oil hole drilling machine described earlier. It is a dedicated machine and a flexible machine. With relatively minor changes, it will drill the oil holes in just about any automotive sized crankshaft. It will not make cylinder blocks or even mill surfaces nor will it drill, ream, and tap miscellaneous holes. If internal combustion engines continue to be built, they will need crankshafts with oil holes.

It is interesting that even today, the Japanese auto companies in Japan and in the U.S. use a significant number of dedicated transfer machines. Some incorporate significant flexibility. Machining center type of machines, where it makes sense, will be in the same lines (agility where it is really useful and not just agility for the sake of agility).

The machining center approach has the advantage that typically several machines are used to perform the same operation. If one is down production continues but on a reduced level. If a transfer machine is down all production in that line stops until the problem is resolved. A transfer machine uses multi-spindle heads where one stroke of a machine axis will produce all the holes in a many bolt hole pattern. The machining center has one feed axis, one tool and one set of spindle bearings, requiring many strokes, to do the same work. It has two other machine axes that must reposition its spindle for every stroke. The safety margin time, when shifting from rapid approach to avoid crashing into the work piece at high speed, is required every stroke of the machine axis. A typically high maintenance tool-changing device is required when changing to another operation during the production process.

Addendum 3:

Managing Production Equipment

The technology levels in American modern production hardware have been elevated to levels beyond the capability of many of those being asked to operate and maintain it. Ineffective training and, more importantly, the serious lack of ownership spirit are important shortcomings. There are cases where the buyer dictates certain specifics for the hardware even with efforts by the seller to discourage too much reach. This is particularly true in equipment control systems, which today nearly always utilize rapidly advancing computer technology.

As a sidelight to this discussion: The American builders are required to follow the letter of the advancing U.S. domestic customer specified computer control. European builders are many times allowed to utilize their own or European standard controls, already proven on earlier generation equipment even for their American customers.

A large part of the time, the new U.S. computer control systems are the first application requiring serious pioneering by the builder. There have been serious problems and subsequent criticism of the builders when the difficulties cause delays even when the control was found to be at fault and probably not really fully developed or ready for the market.

It is true that the special machine tool builders will offer advanced approaches to become more cost effective and competitive over the product life cycle. Once accepted by the buyer, it is incumbent on his organization to provide the training and the environment to effectively manage that equipment on the production site with reasonable support by the supplier.

The production sites in many American automotive and in other manufacturing industries do not enjoy the "turned on" atmosphere discussed earlier in this book. The "life's work" and "reasons we work" logic is foreign, obscured by those with different agendas and the environment is even an adversarial one in some instances. Competitive spirit, passion, and resulting productivity to enhance both the company's and the employee's prosperity and more important, true job security are absent. It can be a blind TGIF environment.

Work rules, established presumably to provide job security to individual labor classifications, wind up hurting them and the entire work force even more. For example, if a machine tool in production goes down due to a failed mechanically actuated limit switch, it can take several different trades people in sequence to restart it. First, to diagnose, then disassemble mechanically, disconnect from fluids (hydraulic, lubrication etc), disconnect electrically, acquire replacement parts, re-connect electrically and hydraulically, re-assemble mechanically, re-set positions etc. re-start and put back into production. Each step requires a different trades person. Any encroachment of one trade on another by supervision or the supplier's service technician invites serious labor problems and further delay. The availability of each of the trades people, in sequence, when actually needed is nearly always not favorable as they are in demand throughout a typically large facility. This procedure will take many times longer than a multi capability "turned on person" would require in a "turned on" organization or a dynamic work team (the competition). The cost in lost production is out of sight.

The same situation observed in a Japanese auto manufacturing plant follows a different routine. As the machine goes down an alarm sounds and a message board highlights the problem machine. Immediately people will be seen running to the site and work to correct the problem begins by whoever arrives first, regardless of the nature of the problem or what his or her specialty is. The fix is accomplished in a small fraction of the time in the first example and production is resumed. Is something wrong with this picture?

Who will win the global cost competitive battle and provide the greatest job security for the individuals involved? For all intents and purposes the individuals in the two different scenes described are competing with each other head-to-head in a global long-term competitive battle for their own job security. Scary, isn't it?

In an American, organized labor plant, as new equipment is put into production, operators and maintenance personnel are trained in its specific requirements. As it happens, if a more desirable job opportunity becomes available some where else in the facility, the freshly trained personnel with seniority may opt to take it negating the training etc. Or, someone whose job has been moved or eliminated because of seniority may bump the freshly trained personnel.

In addition to the obvious potential cost in terms of lost production, this certainly destroys any reason for people to take ownership, as it could be lost at any time anyway. Remember the real definition of job security - the company must succeed against "go for the throat" competitors from around the world or the jobs may no longer exist.

Addendum 4:

History

In the 1960s and 1970s, the American special machine tool industry was flourishing. Six leading companies, four located in the Detroit area and two in Rockford, Illinois, were quite close in size. Their annual sales were in the $60 million range at that time, which would be in the $200 million range in today's dollars. In addition, there were numerous other active companies of real substance many in the Detroit area and others in the Midwest and New England states. These were fine companies, well-known in their circles, respected around the world, and with one exception were closely held. They were prospering, growing, and expanding their technological base.

One of the larger companies, The Cross Company, began a second major operation in Germany and then a third in England and then initiated a 50 percent ownership joint venture in Japan. In the late 1960s, Cross started an industry related computer company of very significant proportions - a visionary concept (computer control of machine tools was in its infancy at that time). A second major manufacturing operation in the U.S. followed.

Another of the larger companies, LaSalle Machine Tool Company, acquired a medium sized machine tool company in Italy, started another through acquisition in Canada and began two other significant operations in the U.S.. Still another, Ingersoll Milling Machine Company, acquired several German machine tool companies over a period of time and another in England while growing and modernizing their domestic operations substantially. In the late 1980s Ingersoll acquired a small machine tool company in Michigan, CM Systems, specializing in crankshaft production machinery. Another of the larger companies, F. Jos. Lamb, added operations in Canada, Germany and in England. A second major operation was added in the U.S. as well. Buhr Machine Tool started an operation in England, expanded their operations in Ann Arbor Michigan substantially and added a significant engineering facility in Detroit..

A Detroit company, another of the larger ones, EX-CELL-O Corporation, itself a division of a larger corporation, acquired a smaller privately held company in Howell Michigan, A.E. Parker, and another larger company in Rockford Illinois, Greenlee Brothers. EX-CELL-O also operated a substantial special machine tool division in Germany and smaller facilities in Canada. H.R. Krueger, one of the smaller companies, whose history goes back to the early automobile days, also acquired an operation in England.

Now, 25 years later, of the mostly closely held companies, one of the larger, Ingersoll, remains American and family-owned (although at the time of this effort the company has been split and the production machine portion sold to a Chinese machine tool company). One other, Lamb, is still American-owned by a large corporation, Unova. There remain a few small closely held companies; however, the rest were closed, absorbed by others or were eventually purchased by foreign competitors, largely German.

Those foreign companies recognized the opportunity to participate in the greatest market opportunities in the world and to take advantage of available talent and technology. A very favorable ongoing currency exchange rate, the DM and then the Euro to the Dollar, made it possible to mix American and German content to provide a lethal competitive posture in the American market.

Several additional foreign builders now have operations in the Detroit area –Italian, German and Spanish companies. All enjoy the favorable currency exchange rate.

The American companies had been designing and building machine tools and entire manufacturing systems for companies around the world, largely for engine and vehicle manufacturers. A number of machines were delivered to Japan and played a roll in the Japanese resurgence. Interestingly, the Japanese bought only one of a kind. They found the way to make later requirements themselves. Later machine generations were, many times, improvements of the purchased machines. Some of these original machines are probably still in operation today.

Several of these companies did a substantial amount of business in the former Soviet Union and other iron curtain countries in that time frame, supplying complete manufacturing systems with appropriate agreement of the governments involved.

Addendum 5:

The Slide

In the late 1960s, The Snyder Corporation, an American old-line, family-held special machine tool company was acquired by Giddings and Lewis, a much larger, publicly owned American standard machine tool company. This company, founded in the early 1900s, had a good record of success, having survived numerous business cycles while maintaining a good reputation. It was fairly diversified, as it served several customer industries, including railroad, pharmaceutical, automotive, heavy vehicle and hydraulic industries. Prior to the acquisition, family members had always managed the company

The acquiring company installed a manager who was an outsider to the special machine tool industry even though there were next generation family members involved in various management positions. The acquired company failed a few years later and was closed, a distinct loss for its owners, employees and especially for its customers.

A very similar set of events was involved with another old-line special machine tool company in Rockford IL., WF and John Barnes, although the purchasing company was a large conglomerate, Babcock and Wilcox. It also acquired another smaller special machine tool company in Rochester Michigan, US Broach and Machine. An up and coming corporate executive from the acquiring company without special machine tool experience managed the division, composed of these two companies. They survived for several years but ultimately, the larger of the two was closed. The other was basically closed as well, although the proprietary property and the name were sold to an employee group who continued to operate on a very small scale.

Also in the late 1960s, another U.S. conglomerate, Bendix Corporation, acquired and eventually consolidated two independent, closely held, American special machine tool companies, Buhr Machine Tool and Michigan Special Machine Tool, and a few years later included a third, Colonial Broach. One of the acquired companies, Buhr Machine Tool, was one of the five larger special machine tool companies at the time. The new company was managed by people from the two acquired companies, but was subject to close scrutiny and the business practices of the parent company. The new company, Bendix Machine Tool Company, performed reasonably well for several years although the competing companies prospered and grew at a faster pace.

The consolidated company, after several years and other changes, was sold to one of the oldest and best known of the other American big five special machine tool companies in the mid-1980s, The Cross Company. Shortly following that acquisition, the acquiring company did it again, this time acquiring another of the big five special machine tool companies of that time, La Salle Machine Tool Company.

The special machine tool division of the acquiring company itself had already undergone significant changes. It had been a public company for a number of years but was managed and controlled by the founding family. It was then merged with a large well-known American standard machine tool company, Kearny and Trecker with a similar family ownership situation, to be known as Cross and Trecker. They were managed as distinctly different businesses under a corporate umbrella.

During the next few years, several changes were made at the corporate level. It eventually led to the appointment of a new chairman from the outside with little knowledge of the businesses he would be responsible for. He in turn selected a president of The Cross Company from the outside. He was a professional manager with no machine tool experience.

This was now a special machine tool company with plants in Canada, England, Germany, four in the U.S., and had an important licensee in Japan. It served customers around the world at the rate of approximately $250 million per year. It was well-known and respected. It now had two large American special machine tool competitors, Lamb Technicon and Ingersoll Milling Machine Company, rather than four. Overall, it was in an excellent market position. The standard machine tool company was a sister division of the same corporation and provided some synergy.

Within two years the company was in serious difficulty. Numerous orders were in serious trouble, technically, financially and for their delivery commitments. This had a very serious negative effect on customer relationships, which adversely affected new orders. The backlog was very low, the company was operating at a loss and numerous projects had serious technical difficulties.

During that difficult period the management at the corporate level was replaced. In turn the management at the special machine tool division was replaced with an experienced special machine tool person from within. Reorganized and working to fix the backlog and warranty problems the company began to show improvements in a year or so and had regained respectability in the eyes of their customers. Within five years, the company was earning at record levels and had a record sized high quality backlog.

This company was performing so well that it was in fact bait for prospective buyers of the corporation including other divisions, which were not doing as well. The corporation was sold to the same standard machine tool company, Giddings and Lewis, as in the first example, this time on a much larger scale.

The acquiring company, managed by an individual with limited standard and no special machine tool experience, chose to split the domestic Cross Company from the European Cross Company pieces. Managers without machine tool experience of any kind would manage each piece of the now fragmented company independently. This was at a time of accelerating globalization by customers and competitors. It also chose to change its name from the one that was the most well-known throughout the global industry. The name of Giddings and Lewis with little connection to the special machine tool customer would be used.

The new American division and new inexperienced president which inherited the large backlog did exceedingly well in the first two years in terms of it's profitability by producing that backlog with essentially the same organization. As the backlog was completed it was not replenished at the same rate.

At about the same time the diminishing backlog became a concern, one of the large domestic auto company customers asked for quotations for equipment to produce major components for a new engine. A distinguishing requirement, surprisingly from a previously "traditionally" oriented customer organization, was "agility" which would require serious pioneering.

In the situation, where the person responsible for the management of the company is not able to assess technical viability and associated risk himself, as the founding entrepreneurs were, he must have the ability to judge the department heads and their recommendations. They should be technically astute, but are less likely to be able to do intelligent risk assessment. Most were in effect trained by their upbringing to depend on the owner/entrepreneur for the tough decisions and were not normally challenged to take risks at that level.

The president accepted what his people had proposed, which was what the customer had asked for, and was awarded a large order for a system to machine aluminum cylinder heads in a way that had not been attempted before.

Another reason that logic may have been distorted in that exercise was that the organization had just processed the large group of important orders very successfully. All, including the president, felt confident in their collective ability when in fact it was the legacy of a different organization. The backlog was already in place, utilized sound practices and pricing based on knowledge of competitors and on market conditions at the time.

Following receipt of that order another domestic automotive customer requested proposals for equipment to produce major engine components. Two different projects were involved. One project would produce cylinder blocks and the other both cylinder blocks and cylinder heads. These would be large systems and were proposed using flexible concepts, but different from the first cylinder head project. Large orders for three lines of equipment were awarded to the same company.

All four systems were over budget, late and shipped without being fully qualified to satisfy customers who were in serious trouble because of it. Worse, they were ultimately shipped back to the builder because they simply couldn't perform as required. Other arrangements had to be made to produce the engine components at great cost to the auto companies and to the machine builder. These were historical precedents and seriously injured the mutual trust aspect of special machine tool acquisition. Both sides shared in the blame.

A leading American special machine tool company, whose genesis included seven once proud and powerful family companies and an untold wealth of knowledge and experience, was mortally wounded. The owner's, the employee's and especially the customer's losses, not just financial, cannot be quantified. It also seriously damaged the rapport that had existed between the two industries for many years.

Some time later the entire corporation, including the remains of the special machine tool company, was sold to the leading German machine tool company. They renamed the special machine tool company to include the original American company name in front of the German name and then changed the name of the entire German special machine tool company the same way – Cross Hüller. In the mid 1980s, EX-

CELL-O corporation, headquartered in Troy, Michigan, closed its machine tool operations in the U.S. and sold its German company to an employee group.

Addendum 6:

Notable American Special Machine Tool Companies

> ➤ No longer producing significant special machine tools under an American name.
> ❖ Producing special machine tools as of December, 2002

- ➤ A.E. Parker and Sons (acquired by EX-CELL-O Corp.)
- ➤ Agnew
- ❖ Ann Arbor Machine Company
- ➤ Apex Corporation
- ➤ Avey (acquired by The Cross Co.)
- ➤ Baker Brothers
- ➤ Bausch
- ➤ Bendix Machine Tool Co. (acquired by The Cross Co.)
- ➤ Biltrite
- ➤ Buhr Machine Tool (combined with Michigan Special as Bendix Machine tool)
- ❖ Cargill Detroit (South East Michigan capacity minimized)
- ➤ The Cross Co.
- ➤ Dearborn Machine
- ➤ EX-CELL-O (the American co,)
- ➤ Fenton Machine Tool (part of LaSalle Machine tool)Co.)
- ➤ Foote-Burt (later Foote-Burt Reynolds)
- ➤ Greenlee Brothers (acquired by EX-CELL-O Corp.)
- ➤ H.R. Krueger
- ❖ ** Ingersoll Milling Machine Company (part of Ingersoll International)
- ❖ Ingersoll CM Systems (part of Ingersoll International – formerly CM Systems)
- ❖ Kingsbury Corporation
- ❖ **Unova (Lamb Technicon - F. Jos. Lamb)**

➢ LaSalle Machine Tool (acquired by Acme Cleveland and then by The Cross Co,)
➢ Michigan Machine (acquired by Lamb - Unova)
➢ Michigan Special Mach.(combined with Buhr Mach. as Bendix Machine Tool))
➢ Natco
➢ Newcor Machine Tool
➢ Place Machine
➢ Producto Machine
➢ R & B Machine Tool (acquired by Lamb - Unova)
➢ Rockford Special
❖ Saginaw Machine Systems
➢ Standard Machine
➢ Standard Machine of New Jersey (acquired by Lamb - Unova)
➢ Snyder Corp. (acquired first by Giddings and Lewis-then by Utica Enterprises)
❖ Tuff Machine
❖ Utica Enterprises
➢ W.F. & John Barnes

38 companies listed
❖ = (9) Continuing operations Dec 02

** In Late 2002 the production machine portion of the business was acquired by a Chinese machine tool company.

Notes:
Highlighted company is the only apparent remaining American special machine tool company capable of supplying major machining systems for the engine and vehicle builders for power train and chassis components.

❖ The others thus denoted either specialize in cylindrical components or very large components or are only able to handle smaller projects.

This is not intended to be a complete history of the subject industry. There are numerous other smaller companies and others that existed earlier in history that are not listed. In addition the companies that are listed, in many cases, underwent other changes and additions too numerous and complex to include.

abundant life
acorns
agile
Agriculture
Albert Speer
Amada
American legacy
application engineering
auction
auto industry
Babcock and Wilcox
benchmarking
Bendix Machine Tool
benevolent cycle
Benjamin Disraeli
Blackbirds
Blue collar
Boeing
Buhr Machine Tool
buy American
Carver
Case
Caterpillar
Celera Genomics Corp
challenging
character
chassis
Christy Borth
Chrysler Corporation
closed shop
closely-held
Coca Cola
cognitive
Comau
communicating

computer
computer aided design
Connecticut Yankee
continuous improvement
conundrum
courage
craft
craftsman
crankshaft
creativity
Cross
Cross and Trecker
Cross Huller
culture
cylinder block
cylinder head
DaimlerChrysler
dead end jobs
dedicated
Dell
democracy
diagnosis
downsizing
drudgery
dynamic value
dynamic work team
Edison
Einstein
Eli Whitney
email
empowers individuals
empowers organizations
enterprise
enterprise personality
enterprise success
entry barrier
equal pay for equal work
EX-CELL-O

exchange rate
F Jos Lamb
Facilitate
failure
farmer
fighter pilot
find a way
fitters
flexibility
florist
Ford
Ford Motor Company
formula
Francis Collins
Franklin D. Roosevelt
free enterprise system
free market
fulfillment
Gates
General Motors Corporation
Germany
Giddings and Lewis
gold rush
gunboat
H R Krueger
hardball
Henry Ford
Honda
huge population
human element
human genome
imagination
indenture
individual liberty
individual power
industrial revolution
industrialized countries
industrials

information age
ingenuity
Ingersoll Milling Machine Company
inquiry
intellectual property
interchangeable
invent
irrational exuberance
isolated population
Italy
James Hoffa
Japan
Jeremiah Wilkinson
Jigs and fixtures
John F. Kennedy
John Quincy Adams
just in time inventory
kaisen
kanban
Karl Benz
Kearny and Trecker
Knudsen
labor
LambTechnicon
LaSalle Machine tool Company
latent
leadership
lean design
lean production
lend lease
liberty
life cycle cost
life expectancy
life quality
life's work
loan guarantees
longevity
Lopez

loyalty
Luddite
machining center
managements preferences and prejudices
manufacturing engineer
market distortions
market share
mass production
maturity
McDonalds
Michael Jordan
Michigan Special Machine tool
Microsoft
mission statement
monopolies
multi-disciplined
musket
mutually dependent
mutually exclusive
mutually supportive
nanotechnology
national medal of technology
natural selection
Neil Armstrong
new economy
Newcomen engine
Nicholas Joseph Cugnot
Nissan
NLRB
no rules environment
numerically controlled machine tool
nurse
old economy
open shop
organized labor
Orville and Wilbur Wright
ownership model
oxymoron

paradigm
passion, definition
Pasteur
Paul Revere
Pearl Harbor
peer pressure
piston
platform
prescription
procurement
products produced
progress
progress curve
progress payments
quality
recognition
retirement
riddle
right to work laws
rust belt
safety net
Salk
Saturn
sealed bid
security
seniority
simultaneous engineering
sixth priority
Snyder Corporation
solution
sour grapes
Soviet Union
special machine tool
stakeholders
Stalin
standard machine tool
standardization
stock market

subsidies
success machine
suckup
surgeon
survival
survival of the fittest
sweet grapes
Swiss watch
target price
tax abatement
team
Teamsters union
tech stocks
technology and medicine
tgif
tgim
The American Revolution
the cause of progress
the human element
Third World countries
Thomas Sowell
tool maker
Toyota
transfer machine
Tupolev
turned off
turned on
Tympanist
U.S. Broach and Machine Company
UAW
unions
Unova
Up From the Apes
USS Constitution (Old Ironsides)
value
values statement
venture capitalist
Vietnamese War

voice mail
Von Braun
W F and John Barnes
Wall Street
white collar
why we work (W3)
wing spar
wisdom
work
working class
world class
WWII
Yamamoto
Yankee
Yankee ingenuity (author's definition)

About the Author

Jim Egbert was born in Northern Wisconsin. He graduated from Henry Ford Trade School in Dearborn, Michigan (high school) in 1951. His first full-time employment as a beginning mechanical draftsman with an engineering firm followed.

Some years later and for a period of seven years, Jim was an equal partner in an engineering company that supported the special machine tool industry. He then joined a large special machine tool company advancing to V.P. Engineering, V.P. Manufacturing, and then V.P. and General Manager of that company's largest division.

In 1987, he became president of one the leading special machine tool companies in the world with four plants in the U.S., one in Canada, one in England, one in Germany and an important licensee in Japan. He retired at the end of 1991. In those years, the company reversed a serious decline and prospered.

He then accepted an opportunity to lead a small division of another major machine tool company. That division prospered, doubled in size and put in place, a foundation of contemporary product technology during the five and a half years prior to his retirement at the end of 1997.

ASPATORE BOOKS

Bulk/CUSTOMIZED BOOK Orders

Aspatore Books offers discount pricing and customization of cover art and text on bulk orders. Customization choices might include but are not limited to: Adding your logo to the spine and cover; Personalizing the book title to include your company's name; Removing specific unwanted content; Adding a letter from your CEO or others; Including an application form or other collateral materials. Companies use Aspatore books for a variety of purposes, including: Customer Acquisition, Customer Retention, Incentives and Premiums, Employee and Management Education. Contact Rachel Pollock at 617.742.8988 or rp@aspatore.com for more information.

LICENSE Content

Aspatore content is often licensed for publications, web sites, newsletters and more. Electronic licenses are also available to make an entire book (or series of books) available for employees and/or customers via your web site. Please contact Jason Edwards at jason@aspatore.com for more information.

ADVERTISE IN C-LEVEL BUSINESS REVIEW

Every quarter, C-Level Business Review reaches thousands of the leading decision makers in the United States. Subscribers to the magazine include C-Level executives (CEO, CFO, CTO, CMO, Partner) from over half the Global 500 and top 200 largest law firms. Please email jonp@aspatore.com for advertising rates and more information.

CORPORATE PUBLISHING GROUP
(AN ASPATORE OWNED COMPANY)

Corporate Publishing Group (CPG) provides companies with on-demand writing and editing resources from the world's best writing teams. Our clients come to CPG for the writing and editing of books, reports, speeches, company. For more information please e-mail rpollock@corporateapublishinggroup.com.

To Order or for More Information From an Aspatore Editor, Visit Us At www.Aspatore.com Or Call Toll Free 1-866-Aspatore (277-2867)

BEST SELLING BOOKS

REFERENCE

Business Travel Bible – Must Have Phone Numbers, Business Resources & Maps

The Golf Course Locator for Business Professionals – Golf Courses Closest to Largest Companies, Law Firms, Cities & Airports

Business Grammar, Style & Usage – Rules for Articulate and Polished Business Writing and Speaking

ExecRecs – Executive Recommendations For The Best Products, Services & Intelligence Executives Use to Excel

Executive Zen – Mental & Physical Health & Happiness for Overworked Business Professionals

The C-Level Test – Business IQ & Personality Test for Professionals of All Levels

The Business Translator-Business Words, Phrases & Customs in Over 65 Languages

MANAGEMENT/CONSULTING

Corporate Ethics – The Business Code of Conduct for Ethical Employees

The Governance Game – Restoring Boardroom Excellence & Credibility in America

Inside the Minds: Leading CEOs – CEOs Reveal the Secrets to Leadership & Profiting in Any Economy

Inside the Minds: The Entrepreneurial Problem Solver – Entrepreneurial Strategies for Identifying Opportunities in the Marketplace

Inside the Minds: Leading Consultants – Industry Leaders Share Their Knowledge on the Art of Consulting

Inside the Minds: Leading Women – What It Takes to Succeed & Have It All in the 21st Century Being There Without Going There: Managing Teams Across Time Zones, Locations and Corporate Boundaries

TECHNOLOGY

Inside the Minds: Leading CTOs – The Secrets to the Art, Science & Future of Technology

Software Product Management – Managing Software Development from Idea to Development to Marketing to Sales

Inside the Minds: The Telecommunications Industry – Leading CEOs Share Their Knowledge on The Future of the Telecommunications Industry

Web 2.0 AC (After Crash) – The Resurgence of the Internet and Technology Economy

Inside the Minds: The Semiconductor Industry – Leading CEOs Share Their Knowledge on the Future of Semiconductors

VENTURE CAPITAL/ENTREPRENEURIAL

Term Sheets & Valuations – A Detailed Look at the Intricacies of Term Sheets & Valuations

Deal Terms – The Finer Points of Deal Structures, Valuations, Term Sheets, Stock Options and Getting Deals Done

Inside the Minds: Leading Deal Makers – Leveraging Your Position and the Art of Deal Making The Art of Deal Making – The Secrets to the Deal Making Process
Hunting Venture Capital – Understanding the VC Process and Capturing an Investment
Inside the Minds: Entrepreneurial Momentum – Gaining Traction for Businesses of All Sizes to Take the Step to the Next Level

LEGAL

Inside the Minds: Privacy Matters – Leading Privacy Visionaries Share Their Knowledge on How Privacy on the Internet Will Affect Everyone
Inside the Minds: Leading Lawyers – Leading Managing Partners Reveal the Secrets to Professional and Personal Success as a Lawyer
Inside the Minds: The Innovative Lawyer – Leading Lawyers Share Their Knowledge on Using Innovation to Gain an Edge
Inside the Minds: Leading Labor Lawyers – Labor Chairs Reveal the Secrets to the Art & Science of Labor Law
Inside the Minds: Leading Litigators – Litigation Chairs Revel the Secrets to the Art & Science of Litigation
Inside the Minds: Leading IP Lawyers – IP Chairs Reveal the Secrets to the Art & Science of IP Law
Inside the Minds: Leading Deal Makers – The Art of Negotiations & Deal Making

FINANCIAL

Inside the Minds: Leading Accountants – The Golden Rules of Accounting & the Future of the Accounting Industry and Profession
Inside the Minds: Leading Investment Bankers – Leading I-Bankers Reveal the Secrets to the Art & Science of Investment Banking
Inside the Minds: The Financial Services Industry – The Future of the Financial Services Industry & Professions
Building a $1,000,000 Nest Egg – 10 Strategies to Gaining Wealth at Any Age
Inside the Minds: The Return of Bullish Investing
Inside the Minds: The Invincibility Shield for Investors

MARKETING/ADVERTISING/PR

Inside the Minds: Leading Marketers–Leading Chief Marketing Officers Reveal the Secrets to Building a Billion Dollar Brand
Inside the Minds: Leading Advertisers – Advertising CEOs Reveal the Tricks of the Advertising Profession
Inside the Minds: The Art of PR – Leading PR CEOs Reveal the Secrets to the Public Relations Profession
Inside the Minds: PR Visionaries – PR CEOS Reveal the Golden Rules
Inside the Minds: Textbook Marketing – The Fundamentals We Should All Know (And Remember) About Marketing

ASPATORE
C-Level Business Intelligence™